Heinz-Uwe Hobohm

Healing Heat

An essay on cancer fever therapy

Immunological basics and practical application with 16 case reports

© 2008-2016 Heinz-Uwe Hobohm. All rights reserved. This publication my not be copied, stored, transformed or changed, either electronically, mechanically, by fotocopy or otherwise, without permission of the author

This essay is a translation from the German version "Heilende Hitze"

Printer and publisher: BoD-Books on Demand, Norderstedt
ISBN 9783734738791

The front picture shows a cancer cell attacked by leukocytes © 123RF GmbH, 61130 Nidderau, Germany

"Are pathogenic and non-pathogenic microorganisms one of Nature's controls of microscopic foci of malignant tissue, and, in making progress in the control of infectious diseases, are we not removing one of Nature's controls of cancer?"

Shear, Journal of the American Medical Association (Vol. 142), 1950. p. 383

Content

Introduction	11
Coley's Toxin	13
Prof.Busch's Experiment	14
The development of Coley's toxin	15
The use of Coley's toxin until 1940	18
Coley's experiments in retrospect	21
Klyuyeva's Toxin	23
Recent experiments using bacterial extracts	25
Spontaneous regressions	29
Medicine: plausible versus factual	33
Tonsillectomy	33
Treatment after thrombosis	35
Cholesterol lowering drugs	36
Rectal cancer	37
Pancreatic cancer resection	37
Mammography	38
Lymph node excision	39
Prostate cancer	39
"The human body has no means against cancer"	41
Evidence based medicine	42
Immunology	43
Tumour antigens	43
Experiments with mice	44
Tumour infiltrating lymphocytes	46
Dendritic cells	47
The innate immune system	48
PRRL	52
Mistletoe therapy	57
Cancer epidemiology	61
PRRL-therapy	69
Past clinical tests with PRRL substances	69

CONTENT

Therapeutic windows for future clinical tests	71
Practical application of a PRRL-mix	75
Exclusion criteria	75
Signs of positive outcome	82
Risk of severe adverse reactions	82
Hyperthermia	**85**
Cancer fever therapy: cases	**87**
Case 1 - B-cell lymphoma	88
Case 2 - Non-Hodgkin lymphoma	90
Case 3 - follikular lymphoma	91
Case 4 - Oesophageal cancer, lung recurrence	92
Case 5 - Malignant melanoma	94
Case 6 - Breast cancer	95
Case 7 - Melanoma lung metastasis	96
Case 8 and 9 - B-cell-lymphoma	98
Case 10 - Merkel cell carcinoma	103
Case 11 - Bilateral breast cancer	104
Case 12 - Breast and mesothelioma cancer	106
Case 13 - Colon carcinoma with hepatic metastasis	107
Case 14 - Metastasing melanoma	108
Case 15 - Melanoma relapse	109
Case 16 - Mamma carcinoma with bone metastases	110
Cases - wrap up	111
Prophylaxis and aftercare	**113**
Diet and cancer	**117**
Vista	**121**
References	**125**

Introduction

Antibiotics and vaccines are considered to be the most important medical achievements in human history. Both keep at bay the threat posed by bacteria, viruses and other pathogens and save millions of lives each year. Chronic bacterial and viral infections are accountable for about one fifth of all cancers.

No wonder that in the eyes of the public, including the vast majority of physicians, an infection and its visible hallmark, fever, for all intents and purposes, are of no value.

But are they?

The human body is generally, both by specialists and patients, regarded as a machine too weak to fight cancer. But this is a misconception.

The human body replaces about 10^7 cells per second (10 million!). If we multiply this number with the number of seconds of an average human lifetime of 80 years, we arrive at the astonishing number of 10^{16} cells a human body produces during its life. Let us assume, we remove all factors from our environment and lifestyle, which are known to promote cancer (bad nutrition, smoking, air pollution, radiation a.s.o.). It has been estimated that the residual incidence of cancer would be about 10%. So one in 10 people would develop clinically relevant cancer during her or his life, rather than 2-4 out of 10, which is the present incidence. The likelihood for a normal cell in a normal human body to give rise to a malignancy therefore would be 1 out of 10^{17}. Robert Weinberg summarized these odds by stating: "In spite of the enormous intrinsic risk of developing cancer, the body must be able to mount highly effective defences that usually succeed in holding off the disease for the 70 or 80 years that most of us spend on this planet." [2].

The major player in this defence is our immune system. Without our immune system we would be overwhelmed by cancer cells within a very short period. The immune system is capable to counteract cancer for long periods, often many years, before a clinically evident neoplasm develops. Second, even in cases of late stage disease so-called spontaneous regressions and cures can occur.

Hundreds of cases of spontaneous regressions have been documented in the medical literature, and we can reasonably assume a much higher number of unrecorded cases. These regressions did not always, but often enough, lead to

INTRODUCTION

definitive cure. Even in cases of late stage, hopeless disease. Hence, in principle the human immune system can destroy even bulky tumours.

Spontaneous regressions continue to be reported in the scientific literature to this day, but not explained. In November 2005 the medical journal "Deutsches Ärzteblatt" states that "with todays knowledge we cannot give recommendations how to best enhance spontaneous regressions".

Careful analysis reveals that most spontaneous regression happened after a feverish infection. This correlation is also valid viewed from the other side. A personal history of feverish infections reduces the likelihood to develop cancer later in life.

Physicians have taken advantage of this strange affiliation more than 100 years ago. They had observed that sometimes after an infection in a cancer patient tumours can become softer or smaller or even disappear completely. Accordingly they treated cancer patients by injecting bacterial extracts many times. On one hand spectacular cures were reported which would be "difficult to achieve now"[112], on the other hand this fever therapy did not operate in every treated patient. Fever therapy became pushed out of the way by radiation and later chemotherapy, which indeed offered faster and better comprehensible results, however, as we know today, those immediate results do not hold in many cases.

This essay tries to align the old observations with modern immunology, to argue, that there is high potential to tie on those old experiments. It will be hardly possible to get approval for bacterial extracts today, or only with considerable expense by establishing a GMP facility and laborious quality controls. However, to the surprise of even many physicians, there are approved drugs on the market which potentially can supersede bacterial extracts by combining them off-label in cancer therapy.

We now understand better what happens in the human body with cancer cells, when confronted with pathogenic substances.

Coley's Toxin

Elisabeth Dashiell was seventeen, when she entered New York Hospital in the autumn of 1890 with bad pain in her hand. The responsible surgeon was a young man in his late twenties named William B. Coley, who had finished his studies less than a year ago. Dashiell's hand had been swollen for several weeks, yet there was no sign of an infection. After first inspection, Coley decided to wait for further developments. By the end of October the symptoms had steadily worsened. She needed morphine to relieve her pain. In November 1890 a biopsy was taken, which sadly showed round cell sarcoma of the hand bone. Sarcoma is a relatively rare form of cancer growing from soft tissue and bone.

Shortly after the biopsy, her arm had to be amputated below the elbow. But her cancer spread with ferocious rapidity. In December a tumour was detected in her right breast, other nodules followed on the left breast within days. In late December a huge tumour occupied the abdomen above the stomach, in January liver and heart began to fail. Elisabeth Dashiell died January 23, 1891.

Picture 1: Elisabeth Dashiell, crop from ref. 1

Medicine in those days had little more to offer cancer patients than amputation and morphine. Coley, shocked by this aggressive course and his own helplessness, spent the hours after his work searching hospital records and literature to learn more about sarcoma cancer. He found about ninety sarcoma case records. About half of these reports contained follow-up histories, with one particularly exciting case of a man called Stein.

Fred Stein, a German immigrant, had been diagnosed with cheek sarcoma in 1884 and had four operations for recurring cancer. He was considered a completely hopeless case. However, in late 1884 Stein developed high fever and was diagnosed with erysipelas, a common post-operative skin disease in those days. To the great surprise of his physicians the tumour disappeared. Stein was discharged from hospital in February 1885.

Five month after Elisabeth Dashiell's tragic death, Coley tracked Stein down in Lower East Side, New York. Stein was examined again and photographed, with no indication of residual cancer.

Stein's surprising recovery drove Coley to dig deeper in Medical Libraries as, if Stein's case was of any significance, there should have been similar cases before.

Coley had studied German in Yale and likely was able to read German medical journals. It is well possible that he stumbled on another interesting report which was published more than two decades earlier, in 1868 in the "Berliner Klinische Wochenschrift". This publication, together with Stein's case, may have stimulated him to undertake his pioneering work in the years to come.

Prof.Busch's Experiment

On page 137 of the German clinical journal "Berliner Klinische Wochenschrift", in the issue of March 23, 1868, a brave man named Professor Wilhelm Busch reported on an incident that had happened half a year earlier, in November 1867[3].

Apparently, he had observed in cases of face and neck sarcoma, that an erysipelas infection sometimes led to a "resorption" of tumour mass: a similar observation Coley made with Stein two decades later.

Picture 2: Berliner klinische Wochenschrift 23. March 1868. Headline of Prof. Busch's article

Free from any objection by ethics-committees, he decided to test whether this observation could be exploited to patients benefit.

In summer 1867, by coincidence, both a patient with a "Kopfrose", a mild erysipelas infection appearing after an injury, and a 19-year old girl with a huge sarcoma of the neck, entered Busch's clinic at the same time.

The sarcoma had increased within five month to the size of a child's head, such that her breathing was compromised and she could not completely close one of her eyes.

It should be noted, that in the pre-antibiotics days erysipelas was one of the leading causes of death from postoperative infections in hospitals. Outside the hospital, *Streptococcus pyogenes*, the pathogen causing erysipelas, killed about 10% of infected patients. It was a dangerous bug, and its clinical application was justified only in extremis. In this case, however, the situation left no doubt. The young lady carried a death sentence, her situation could not be more desperate.

Prof.Busch burned a small area of skin over the girls tumour and onto this placed a cotton pad, which was taken from the erysipelas patient. She developed high fever of 40 degrees Celsius, and the surrounding skin showed signs of an erysipelas infection. To everyone's great surprise, the tumour, which had been tight and dense before, softened and shrank with impressive speed. Within two weeks the tumour reduced to the size of a small apple. She could close her eyes and breeze freely.

Picture 2: William Coley about 1888, picture from ref.1

Unfortunately, the young lady developed circulatory problems, and many efforts had to be taken to strengthen her weak condition. With the disappearance of the "Kopfrose" or "Rothlauf" - the inflamed skin - the tumour grew again and reached its prior size one month after the treatment started. She left the clinic with unknown fate.

The development of Coley's toxin

By his literature investigations, Coley had found more than forty cases of "disappearance of malignancies during an erysipelas attack". He also came across another medical pioneer, Friedrich Fehleisen in Germany, who for the first time had used cultured bacteria, that is, bacteria grown in a culture dish rather than taken from a patient directly, to treat a cancer patient[4].

In his first publication Fehleisen reports that he induced "Rothlauf" (erysipelas) in a 58-year old female patient with multiple skin sarcoma. He was able, by infecting a large tumour mass directly, to induce high fever, a reduction in size of several smaller tumours and initial swelling and later shrinkage of the injected tumour (the long-term outcome of the experiment was not reported). Fehleisen later repeated his method with more patients, but had mixed successes and failures. His work was discontinued.

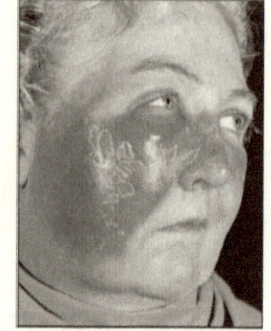

Picture 4: Facial erysipelas infection. From Roche Lexikon Medizin

Still, Coley decided to try again what Prof. Busch had pioneered and Fehleisen elaborated.

In April 1891 an Italian immigrant, Mr. Zola, presented at New York Hospital with a large sarcoma of the neck and an egg-sized metastasis in the right tonsil. He had been operated twice before at other places but was still in hopeless condition. He could hardly talk or swallow liquid and was unable to eat solid food. His life expectancy was, at the very most, a few month, if not much less. In any case, there was nothing to lose.

Bacteria were prepared by colleagues at the College of Physicians and Surgeons, now part of Columbia University. Bacteria were either grown in a gelatin-like culture or in beef broth and applied by rubbing the cultures into small wounds or by injection. Since erysipelas was so hazardous, the hospital was reluctant to host the experiment, so it was done in a private apartment. Over a few weeks, three applications were performed, with little success. Zola's temperature rose only slightly, and he showed no sign of full blown infection. So Coley tried a fresh preparation and a larger dose. Within hours Zola developed severe chills,

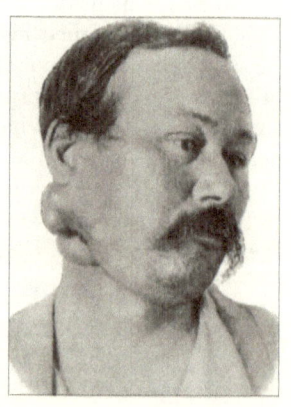

Picture 5: Monsignore Zola before treatment. Picture from ref.1

headache and vomiting. His temperature, however, did not exceed 39 degree Celsius, less than what one could expect from a full-blown erysipelas infection. But the tumour diminished in size, as did the tonsil tumour. About one month after treatment started, Zola could eat food again. His general condition improved - a cause of great excitement.

Yet, despite on-going efforts, Coley was not able to induce a full-blown infection.

The leading bacteriologist at that time was Robert Koch in Berlin, Germany. Via a friend, who visited Koch's laboratory in summer 1891, Coley asked Koch for a fresh bacteria culture, preferably grown from a patient with full-blown erysipelas infection. Koch was willing to help, and in October 1891 Coley started work again at Zola's bedside. This time he was able to induce a hefty infection within one hour of vaccination. Zola's temperature rose above 40 degree Celsius with nausea, vomiting and severe pain, the infection almost killed him.

At the end of two weeks the neck tumour was gone.

The residual tonsil tumour was still present but halted growth permanently. Zola was in excellent health when Coley saw him four years later in October 1895 for the last time.

In 1899, more than eight years after Coley's treatment, Monsignore Zola died in his homeland Italy from unknown cause.

During the following two years Coley attempted to infect twelve inoperable cancer patients. He failed to induce a full-blown infection in four carcinoma patients and succeeded in eight sarcoma patients. All of the these eight responded. Six showed partial tumour remissions and two showed full remission with even metastases vanishing.

Two of the twelve treated patients died from the infection. In corollary, Coley found what Friedrich Fehleisen had found some years ago: the live bacterium could be very dangerous and very beneficial at the same time.

This was the point where Coley went a step further than his predecessors Busch and Fehleisen. He abandoned living cultures and turned towards what today we would call a bacterial extract.

Together with Alexander Lambert, later president of the American medical association, Coley inactivated the microbes either by heating or by filtering through a porcelain-filter, such that in the latter case only culture medium and molecules produced by the bacteria, but not bacteria themselves, could pass the filter.

These preparations were tested on four patients, with modest fever-inducing effects and modest temporary effects on the tumours. The preparations, though no longer dangerous, were too weak.

By the end of 1892 a French doctor named G.H.Roger had published his observation that the virulence of erysipelas bacteria was increased when it was grown in the presence of another bacterium, which at that time was named *Bacillus prodigiosus* and is nowadays called *Serratia marcescens*[5]. This bacterium is a mild pathogen involved in nosocomial, eye and urinary infections.

In January 1893 Coley for the first time administered one variant preparation of what until today collectively is termed "Coley's toxins". It was a porcelain filtered combined culture of *Streptococcus pyogenes* and *Serratia marcescens* bacteria. The patient was a sixteen years old boy with a large inoperable abdominal malignant sarcoma. Over a ten week period, an increasing dosage of the toxin was applied, two to three times per week, in order to slowly approach the optimal concentration of toxin that the individual patient would react to[i].

Finally, the optimal dosage was reached and the boy's symptoms mimicked those of a full blown erysipelas: chills, headache, fever, local redness and swelling at injection site. And the tumour responded, it shrank by 80 percent. The treatment was stopped in May 1893.

Coley kept in touch with his patient, who remained to be healthy over more than 20 years. The boy had been cured of an advanced, inoperable cancer.

Coley treated another five patients during 1893. No result was as promising as the first one.

The use of Coley's toxin until 1940

Coley published the results of his experiments in the American Journal of Medical Science under the title "The treatment of malignant tumours by

---

i   It would likely save us thousands of un-necessary deaths from medical drug side-effects (NEJM 342, 2000, 1123-1125) if we nowadays would abide by this principal of individual dosing – which is demanded by our different equipment with genes reacting on drugs - more readily.

repeated inoculations of erysipelas: with a report of ten original cases" in 1893[6]. The report stirred considerable excitement.

For a while.

At the beginning of the twentieth century, X-ray treatment was invented. This new modality caught almost the total attention of the entire oncological medical community due to its immediately visible effects. One could now, it seemed, burn away tumours. Coley's method drifted out of mainstreams sight.

Nevertheless, some physicians tried to test Coley's treatment.

Nicholas Senn of Rush Medical College in Chicago reported uniform failure of the method in 1895. William Keen, a surgeon in Philadelphia, failed to obtain a response in seven patients..Caulkins of Watertown, New York reported a large number of successes[i], as did Matagne from Belgium, who used to prepare his own fresh extracts rather than using extracts made by others, which might have lost potency during transport. Matagne published his observations in French and Belgium journals of low distribution frequency and impact (see[7] for references) which have never been recognised appropriately in the medical literature.

Two stubborn surgeons at the US-Marine-Hospital in Stapleton, New York named S.L.Christian and L.A.Palmer reported a spectacular cure in 1928[8]. US-Marine Captain G.B. was 31 and suffering bone sarcoma which was treated by an above-knee amputation. Due to residual disease, daily (!) injections of "Coley's fluid" were started January 5, 1926 with stepwise increasing dosage until February 20. "On February 20 they were discontinued because of extreme weakness of the patient." On March 27 injections directly into the tumour mass on the stump were started again using a low dose and continued daily with increasing dose until April 7th. "During May and June the patient grew steadily worse with metastatic growths appearing in many parts of the body, among which was considerable involvement of the right clavicle and multiple tumours in the scalp, cranial bones and cervical vertebrae." On August 5th, 1926 "Coley's fluid was again begun with a dosage of two minims and increased a minim a day until he was receiving 17 minims at a dose. The dose was held at this point until September 4th. By this time improvement was marked.".

---

i   Helen Coley-Nauts (Acta MedScand 1953;145:5-102) reports that Caulkins achieved 80% five year survival in using this technique over a 32 year period" but fails to give a reference. His records allegedly can be found at Cancer Research Institute in New York

Another series of injections commenced on September 19 and continued for three weeks. At this point, a total of about 15 weeks of daily injections had been accumulated. "On November 22, the general condition of the patient was excellent." But the two doctors were not yet confident. The patient was treated again for 5 weeks starting on February 1927 with injections in three-day intervals and again from October to December 1927 with injections in three-day intervals. "The patient was last examined January 9, 1928, at which time there was no evidence of disease present."

This case study was discussed during a conference at the memorial Hospital, New York in December 1927, where Coley stated: "I believe this is one of the most remarkable cases of malignant tumour of the long bones that has ever been published, and I am quite willing to admit that, had the patient been under my care, he would probably not be alive today. At the first place, I am almost certain that I should not have continued the treatment after three month when not only no improvement had been noticed but marked increase had taken place in the metastatic tumours and especially in the recurrent tumour of the stump (an increase from 17 to 31 inches). In the second place, I am quite sure that I should not have dared to increase the dose to such a large amount (30 minims)."

Coley, throughout his career of over more than four decades treated hundreds of patients. He never achieved a clear cut, uniform result. Some patients responded, some were cured, some did not respond[i].

In May 1934 a symposium on "Ewing's sarcoma" took place at Memorial Hospital. Ewing's sarcoma is a rare and usually lethal form of sarcoma, amputation was the method of choice. Cure could often not be achieved due to metastases which had already spread before operation. At this meeting, Coley discussed 44 cases of Ewing's sarcoma, that had been accepted into the "Bone Sarcoma Registry". This central clearing house was founded in 1920 to record pathology slides of sarcoma cases, to ensure correct diagnosis, to register treatment and, if possible, to follow-up patients over time to determine the success of treatments.

The success of cancer treatments, to the present day, is most often judged by the so-called "5-year-survival rate", that is the fraction of treated patients living more than five years after diagnosis without relapse.

---

i   Such unpredictabiliy unsettles clinicians, and one can get the impression that sometimes a method is preferred just because the treatment is short and its short-term outcome can be anticipated with some certainty.

Twelve out of 44 patients had been treated by other physicians with radiation, none had survived five years. 32 out of 44 patients had been treated by Coley using the bacterial extract, twelve of those 32 remained free of disease for more than five years.

A sample of 44 patients is small. Nowadays this would be equivalent to a "Phase-I"-trial, the first level of clinical tests with a few to a few dozen patients to evaluate safety, feasibility and response. In case of a positive outcome a Phase-I-trial should be substantiated by a Phase-II-trial with a hundred or more patients, followed by, upon promising outcome again, a Phase-III-trial with some thousand patients or more. Hence, this result was of lowest order of statistical significance, in particular, since the cases have been selected by Coley retrospectively - nowadays lege artis of cause would be to select the patients for the trial and treat them afterwards.

Still, the result – a five year survival rate of zero percent after radiation and 38 percent after Coley's method - is so puzzling that it definitely would have required deeper scrutiny.

Coley's experiments in retrospect

In the year 1953, in a report of almost 100 pages in the journal "Acta Medica Scandinavica", Coley's daughter Helen Coley-Nauts re-examined in detail the clinical cases treated by her father[7]. Later, in the 1980s and 1990s, she summarized the results of Coley and his contemporaries in several publications[9].

Picture 6: Helen Coley-Nauts, picture from http://www.cancerresearch.org/nautsob.html

This was not easy. Coley, undoubtedly a man of determination and impetus, was not a great scientist, neither by training nor by talent. He believed, throughout his life, that microbes cause cancer, his patient records were incomprehensive, he treated different patients over different periods of time, his bacterial extracts were, over time, prepared in different ways. Helen Coley-Nauts found that he had used 15 different preparations. Eleven of them were what she called, "not potent enough". The more potent ones were those based on heat sterilized bacteria rather than culture filtrates. Immunologically, this makes a lot of sense, as we will see later. So, Coley's records were a mess, and I will restrict myself to the data presented by Helen Coley-Nauts and Charles Starnes, a trained scientist, who revisited Coley's cases in 1994 in an interesting and comprehensive review of 35 pages including a long list of literature references covering the work of Coley and contemporary physicians[10]. This review, unfortunately, was not well recognized in the medical literature, almost certainly because it was published in a low-impact journal[i].

Coley had treated several hundreds of patients by the time he died in 1936, but many of them were treated both with bacterial extract and by radiation, many also by surgery. If one wants to estimate the overall success of extract, the analysis must be restricted to those patients who were inoperable and treated by toxin alone but not radiation. Starnes identified 170 of those (121 with some form of sarcoma, 43 with carcinoma and myeloma, 6 with melanoma). The remission rate was 64% and the 5-year survival rate was more than 44% (for some patients no record of more than 5 years could be obtained, they were, as medical parlance would say, "lost on follow-up").

A five-year survival after an inoperable cancer disease is assumed to be a cure. A cure rate of more than 44% is an extraordinary success. If one considers that all these were late-stage cancers with a worst possible prognosis, this rate is nothing less than miraculous.

According to Coley-Nauts and Starnes analyses published in the 1990s, treatment success correlated positively with length of treatment and with height of body temperature induced by the toxin. Starnes pointed out that the whole picture suggests that particular cancers such as sarcoma, but also some forms of carcinoma and melanoma, collectively called "cancers of mesodermal origin", might react better than other forms of cancer to the bacterial toxins - a hypothesis so far. These parameters - treatment length and fever height, but also possible selection of particularly favourable forms of cancer - have not been considered appropriately in the follow-up studies on Coley's toxin, which were done in the second half of the twentieth century.

---

i   For each scientific journal an "impact factor" is calculated from the number of articles cited in other journals during the last 5 years. Journals are ranked based on the impact factor, with higher impact factors for better journals.

Klyuyeva's Toxin

In the years 1946 to 1953 Klyuyeva and co-workers in Russia treated cancer patients using Trypanosoma extracts. These experiments were hardly considered by contemporaries and are long forgotten, but they are worth of reconsideration.

Trypanosoma are parasites causing Chaga's disease, which was and is endemic in Latin America. Trypanosoma have nothing much in common with Streptococci bacteria used by Coley, except that both organisms, like most pathogens, produce substances called PRRL which can cause rapid red alert in the immune system. We will come back to this later.

Klyuyeva's experiments were inspired both by Coley's experiments and observations like the one cited in the introduction of a book published by Klyuyeva in 1957 in Russian language and translated 1963 under the title "Biotherapy of malignant tumours"[11]:

"...according to the Cancer Centre in Sao Paulo (Brazil), among tens of thousands of cancer patients only two gave a positive Machado reaction [typical for patients with past or ongoing trypanosoma infection], whereas among the remaining population the number suffering from this infection varies from 10 to 20 percent."

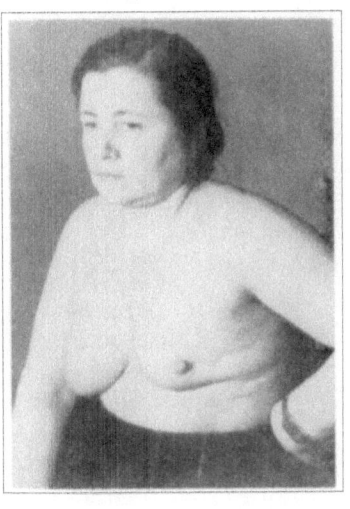

Picture 7: Patient L. (Nr.19) treated by Klyuyeva and coworkers in Russia between 1946 and 1949

The authors do not mention which year this observation refers to, and the numbers "tens of thousands", and "two" ring a little bell of scepticism in a scientist's belly because their level of precision deviates much. But even if these numbers are not exactly correct, the qualitative observation probably is: most cancer patients were not infected by Trypanosoma before and infected people usually do not get cancer. That is puzzling.

COLEY'S TOXIN

Klyuyeva and Roskin, in their book[i] describe 32 patients with cancer treated using *Trypanosoma cruzi* extracts, case by case presenting pictures, lab data and disease sentence.

21 of these patients had lip or breast cancer, several had palpable lymph nodes indicative of progressed disease, all showed no sign of residual disease after treatment at the end of the observation period, with 11 patients monitored more than 5 years.

The book does not contain statistics. Treatment failures were not reported, there was no control group. Extracts were prepared in more than seven different forms, injection site and extract dosage varied.

This sort of anecdotal reports is, for many good reasons, not well respected in the medical community. However, diagnosis and follow up probably were done correctly, each case is described including histological and patient pictures. Let's examine case number 19, the most advanced case of breast cancer in the book.

Patient L., aged 34 years, was diagnosed with breast cancer in Leningrad in the year 1946. Radical surgery was recommended but not accepted by the patient. She received several X-rays, but than refused further treatment. Tumour growth continued, her general condition deteriorated. When the treatment with Trypanosoma injections commenced, the firm, uneven tumour had reached a size of 8 x 6 cm, and four lymph nodes were palpable. The patient received 51 daily injections until October 1946. Usually, Klyuyeva and co-workers used intra-muscular injections into the buttock, but in this case in addition systemic injections and 9 intra-tumoural injections were given. After these injections the tumour size had decreased to 4 x 3.5 cm, only one of four lymph nodes was still palpable. In October 1946 she went back home to Kiew, where she was re-examined. Her physicians decided that now a non-radical surgery might be applicable, and the neoplasm was resected. Histopathological analysis revealed "only isolated cancer cells in stages of atrophy". A second round of 40 injections was started in May 1947 for prophylactic reasons (!), a third round of 27 injections in February 1949. In February 1956 examinations in the hospital in Kiew "showed the absence of any signs of recurrence or metastases".

Klyuyeva and contemporaries presumed that the effects triggered by *Trypanosoma* extracts are caused by a hypothetical anti-cancerous substance

i   http://www.fevertherapy.eu/references/

which they called "cancer antibiotic". But antibiotics usually work after a few applications, while Klyuyeva injected the extract daily or near-daily over many weeks, in some cases more than 300 times per patient. Thus, it must have been obvious to them that the comparison with typical antibiotics is somewhat out of kilter.

As Coley did, Klyuyeva and co-workers increased dosage up to a level which they felt was not yet toxic. Febrile reactions after shots were observed, but were obviously not deemed important by them.

We can safely assume that their method did not lead to remissions or cures in all cases, otherwise it would have meant the solution to cancer. But they were clearly on the track of something important. However, in the 1950ies chemotherapy was invented and adopted worldwide with enthusiasm, replacing older therapy methods.

Recent experiments using bacterial extracts

It took decades from the emergence of radiotherapy at the beginning of the twentieth century until it was acknowledged that it could not cure cancer in a majority of cases. X-rays can kill cancer cells and induce later cancer at the same time, so one has to compromise the irradiation dosage. Although this is well known for more than hundred years, even in the 1980ies a batty chief physician in Hamburg managed to severely harm many cancer patients by over-dosing, following his simplistic strategy "much helps much".

The same happened when chemotherapy was developed in the middle of the century. Today we have to recognise that surgery contributes most to the vast majority of cancer cures, while radiation and chemotherapy alone, although clearly prolonging survival within the five-year window, are responsible for only a small percentage of all cures (survival of more than five years). Nevertheless, hypes and hopes fuelled by both methods for a long time and overshadowed alternative approaches.

There were only a few uncoordinated attempts to revive Coley's ideas in the second half of the twentieth century[10]. The bacterial extracts used in these studies were commercial preparations called MBV (Bayer) or Vaccineurin (Suedpharma, Munich) similar to Coley's extracts. Vaccineurin was approved for "neuralgic pain" and production was stopped in the late 1980ies. It was used off-label in many private clinics in Germany in the 1960ies and 1970ies for cancer treatment. Regrettably, results have never been summarized and published.

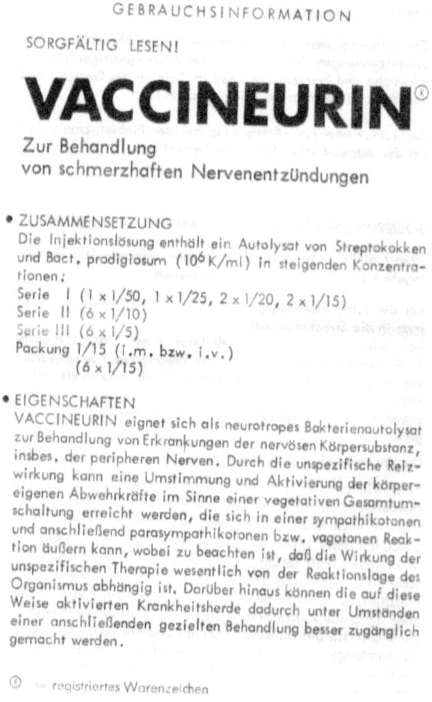
GEBRAUCHSINFORMATION

SORGFÄLTIG LESEN!

VACCINEURIN®

Zur Behandlung
von schmerzhaften Nervenentzündungen

• ZUSAMMENSETZUNG
Die Injektionslösung enthält ein Autolysat von Streptokokken und Bact. prodigiosum (10^6 K/ml) in steigenden Konzentrationen:
Serie I (1 x 1/50, 1 x 1/25, 2 x 1/20, 2 x 1/15)
Serie II (6 x 1/10)
Serie III (6 x 1/5)
Packung 1/15 (i.m. bzw. i.v.)
(6 x 1/15)

• EIGENSCHAFTEN
VACCINEURIN eignet sich als neurotropes Bakterienautolysat zur Behandlung von Erkrankungen der nervösen Körpersubstanz, insbes. der peripheren Nerven. Durch die unspezifische Reizwirkung kann eine Umstimmung und Aktivierung der körpereigenen Abwehrkräfte im Sinne einer vegetativen Gesamtumschaltung erreicht werden, die sich in einer sympathikotonen und anschließend parasympathikotonen bzw. vagotonen Reaktion äußern kann, wobei zu beachten ist, daß die Wirkung der unspezifischen Therapie wesentlich von der Reaktionslage des Organismus abhängig ist. Darüber hinaus können die auf diese Weise aktivierten Krankheitsherde dadurch unter Umständen einer anschließenden gezielten Behandlung besser zugänglich gemacht werden.

® = registriertes Warenzeichen

Picture 8: Vaccineurin, first page of instruction leaflet. Vaccineurin was originally approved for "painful nerve inflammations". Similar to Coley's toxin it was prepared from heat sterilized "streptococci and Bakt.prodigiosum" and used off-label for cancer fever therapy in German private clinics between 1960 and 2000 in thousands of cases. Results from these sub-academic treatments were never published.

MBV and Vaccineurin contained bacterial extracts similar to, but not identical with Coley's preparation. A more important difference in this second generation of studies was that a majority of the patients in these studies had been pre-treated by chemotherapy and / or radiation; both methods have immune-compromising effects which might interfere with a subsequent immunotherapy like Coley's. The experimenters believed probably, as Coley did, that these extracts contain an anti-cancerous substance. As we know today, this was right and wrong at the same time. These substances exist, but they require a fully functional immune system to deploy their beneficial effects.

In many cases, treatment was much shorter than in Coley's time. In many cases, no dosage adjustment was undertaken to exceed a certain threshold body temperature, as was - not always, but usually - done by Coley. The results were mixed: several remissions, even long lasting, several failures. No clear picture emerged [10].

The method was not accepted.

Coley's "miraculous" result of curing almost a half late-stage cancer patients treated exclusively by bacterial extracts remains unexplained. The prospect of curing even at an even higher rate by applying the method to patients with less advanced cancers remains unexploited. Up to the time of writing this essay, a modern, well focused, well controlled study of cancer treatment by bacterial extracts respecting the lessons from the retrospective analysis of Coley's and others cases has not been undertaken.

How can this be ? How can, what a layman understands as a method which might offer an opportunity window in some cases of cancer, be completely disregarded by the specialists, who have the skills and responsibility to implement all possibilities for helping their patients ?

Because specialists implement only what they understand.

Acupuncture, homeopathy, Chinese medicine are all examples of methods known to help in particular cases, but they are understood neither in detail nor superficially, so there is a deep reluctance to implement and even test them. The same holds for Coley's method.

A further problem is that all these methods cannot be patented, there is no big business in sight, so who is to pay enormously expensive clinical trials ?

Thus a combination of a lack of understanding and absence of interest from big pharma are insurmountable obstacles to evaluate a new method. With better understanding, though, there might be a chance to initiate at least some small animal and later clinical tests.

Let's try to improve our understanding. Is there some magic anti-cancerous substance produced by *Streptococcus pyogenes*, as Coley and co-workers believed, which can help at least a proportion of patients ?

In many cases in the history of science, understanding has been gained once a broader view has been captured, possibly by leaving the horizon of the discipline and by borrowing insight from adjacent scientific disciplines. Let us see if we can get some more understanding what happens to cancer patients exposed to live rather than killed bacteria or their extracts. Let's have a look into immunology, epidemiology and case studies on spontaneous regression from cancer. Here we go.

Spontaneous regressions

Spontaneous regression or remission has been defined as "partial or complete disappearance of a malignant tumour in the absence of all treatment or in the presence of therapy which is considered inadequate to exert a significant influence on neoplastic disease"[12]. Actually, the existence of the mere phenomenon may come as a surprise to many readers – can cancer really heal spontaneously ?

Sure enough. There are about 1000 case studies published in the medical literature. over the last century, on spontaneous regression from cancer, but we can be certain that this is the tip of an iceberg of numerous unreported cases. Cases will not be observed when the patients after improvement of their condition did not show up in the clinic any more. Observed cases will not be reported, when the responsible physician either did not consider the case relevant or was unable to make sense out of it or could not rule out effective prior treatment or was not literate enough or did not take the effort to write a report. In other cases false treatment success was deduced incorrectly. Rohdenburg, in a classical article from 1918[13] wrote: "one or another type of medical treatment, the particular therapeutic procedure which is supposed to have brought about the recession in each case being at once hailed as a 'cure' only to give rise to failure and disappointment when tried on a large group of cases. It is probable, therefore, that the isolated cases referred to are really spontaneous recessions and not therapeutic cures".

The phenomenon is rare, though – it was estimated to happen in about 1/100000 to 1/million cancer cases[12], with some forms of cancer such as melanoma being more likely to resolve than others – but that it exists there is no doubt.

As indicated by the definition above, spontaneous regression does not always mean cure – sometimes the regression lasts only for a limited time – but in a significant number of cases durable regression progressed to permanent cure.

Spontaneous regressions have been observed by watchful physicians again and again but ignored at large. Spontaneous regression from cancer cannot, as yet, be explained[i], so the phenomenon is disregarded in the medical literature and hardly investigated. In fact, the mere term "spontaneous" pinpoints our cluelessness since the shrinkage of a large tumour or the evaporation of a leukaemia must have some cause. It does not happen out of the blue.

Is there anything that all these cases might have in common ? Surprisingly, in a considerable fraction - at least in about 30% of cases and likely more - the spontaneous regression could be traced to a preceding infection with fever[14]. For instance, Diamond and Luhby reported 26 spontaneous remissions in a cohort of 300 cases of childhood leukaemia, 21/26 (80%) were accompanied by infection[15]. Stephenson et al investigated 224 cases of spontaneous regression and reported that in 62/224 cases (28%) regression was preceded by either an infection or a persistent temperature elevation. They published this analysis in 1971[16] , where they noted this coincidence among other observations without emphasizing it. It should be noted already at this point, that most likely more infections had occurred, but were not reported in the original articles, such that Stephenson could not count them, because the authors did not imagine any connection.

In many cases, *Streptococcus pyogenes*, the pathogen leading to Erysipelas, was responsible, but not exclusively. Other types of bacteria were involved as well[ii].

So, in a large proportion of cases of spontaneous regression an infection, often a severe one, had occurred shortly before the regression commenced. Without doubt, thus, there is an association in time between these two events in a large proportion of cases.

Non-scientists and even many scientists often tend to understand statistical associations in time as causal connections. However, often, a rigorous sta-

---

i   In an article in the prestigious journal „Proceedings of the National Academy of Sciences" (PNAS 2003;100:6682-7) in 2003 Cui et al. wrote: „Despite efforts over many decades, the mechanism(s) of spontaneous regression in humans and animals has remained elusive."

ii   In 1998, Maurer and Kölmel (Onkologie 1998;21:14-8) reported 68 cases of spontaneous melanoma regression (16 partial, 52 complete). Among these, 21 cases of prior febrile infections were determined (erysipelas 9, postoperative abscess 8, other 4) . Rohdenburg (1918) reports that „recession has occurred after small-pox, pneumonia, malaria and acute tuberculosis."

tistical analysis disproves such inference. For instance, if in a group of 23 people two people have the same birthday, most of us will believe that this can hardly be accidental and start to look for a reason how this could happen. Yet, the likelihood that this happens by chance is more than 50%.

Having said that, let us still assume, as a working hypothesis for which there is up to this point neither proof nor strong evidence, that infections can sometimes trigger spontaneous regression from cancer, just for the sake of working deeper into the subject. Let us bear in mind the risk of invalid conclusion about causal connection between two consecutive events.

Thus, several types of infections have been reported to precede a spontaneous regression. In theory it could be that all these pathogens produce, as Coley believed for *Streptococcus pyogenes*, some magic anti-cancerous substance. But since even malaria was reported to trigger regression, a disease which is caused by plasmodia, another sort of pathogen distinct from viruses and bacteria, it is rather unlikely that the very same anti-cancerous substance is produced by all of them, given the evolutionary divergence of the pathogens involved.

If it is not the invader, could it be the host ? Had, for instance, all these individuals something in common, maybe a particular gene, such that the infection led to regression in them only but not in other cancer patients ?

This is indeed a possibility which cannot be ruled out. On the other hand, this cannot be investigated retrospectively, since the reported cases of spontaneous regression were distributed over the twentieth century. It would, however, be an interesting test which since a couple of years can be done by use of so-called DNA-chips. With gene profiling technology it is possible to compare the genes of a patients in whom cancer surprisingly regressed with those patients in whom it did not. Until further investigations have been carried out, we will have to leave this explanation in the realm of speculation.

An alternative explanation for spontaneous regression after infection is that the subsequent immune reactions triggered by any infection could have side effects of a yet unknown nature that are mischievous for the tumour and beneficial for the patient.

But can the immune system attack tumours at all ? Millions of people dying from cancer each year seem to prove the opposite.

Medicine: plausible versus factual

In 1956 a striking correlation between the invasion of a tumour by immune cells and survival was reported for gastric cancer: the more immune cells present in or around the tumour[i], the longer the average survival of the patient[17]. The invasion can be massive: in laboratory animals up to 40 percent of all cells in a tumour were "infiltrating macrophages", immune cells[18].

This is a very interesting observation, and one expects that its publication should have spawned vigorous research activity about the immune system and its role in the development of cancer.

However, nothing happened.

The 1950s was the decade of the development of chemotherapy. Hypes and hopes were sprouting out, the conquest of cancer over the next few years was envisaged. The publication was ignored, not least because it was published in a low impact journal with a limited distribution. Further, it was published in a century in which a majority of clinical oncologists would simply deny any healing capacity of the human body with respect to cancer.

At this point it might be instrumental to sidestep into the art of present day medicine to consider that medical profession is stacked with prejudice. This is not to argue the fact that modern medicine saves thousands of thousands of lives each year, but to question the common belief that the progress of medicine and medical decision making is always based on solid facts and rigorous inference. The art of medicine sadly holds more and bigger holes of preoccupation than other sciences like physics, chemistry, biology or engineering.

Tonsillectomy

Tonsillectomy is advised by physicians to help children - and to a lesser extent adults - with sore throats. It has been one of the most common operations over more than 70 years.

---

i   Earlier observations of this kind date back to at least 1921 (MacCarty WC, Mahle AE. Relation of differentiation and lymphocytic infiltration to postoperative longevity in gastric carcinoma. J Lab Clin Med 6: 473-480)

MEDICINE: PLAUSIBLE VERSUS FACTUAL

Despite its wide application, rigorous criteria for its use do not exist, and the recommendation for tonsillectomy is based on a couple of fuzzy assumptions.

Unproven assumption number one is that tonsils are dispensable. Although it appears that people can live well without tonsils, the possibility has not been ruled out that tonsil infections might be an important part of the immune system as a learning sub-system of the body, with positive effects revealing only later in life.

Underlying assumption number two is that the removal of an organ of the respiratory tract where infections are often located, leads to a reduction in frequency or risk of those infections. This has never been proven.

Underlying assumption three is that tonsillectomy is a riskless surgery. That is not really true. Any anaesthesia carries a risk: about 1:14000 - 1:20000[19] for fatal complications.

Underlying assumption (of parents) number four is that if a physician recommends tonsillectomy for a child he will have good reason to do so. This is contentious.

In one classic study involving 1000 school children in the US, where two thirds had already lost their tonsils, the remaining third was inspected by a group of school doctors, who recommended about half of this group, i.e. 1/6th of the entire cohort of 1000 for tonsillectomy. The other half was sent to a second group of doctors, who recommended surgery for about half of them. These children had been rejected for surgery twice before and were recommended for surgery now. The remaining children were than sent to a third group of doctors, who recommended surgery to about half of them. At this point only 65 of the original 1000 had not been recommended for operation.

Although this study dates back to pre-war times, even today, when the use of antibiotics enables respiratory infections to be managed less dramatically compared to pre-antibiotic days, tonsillectomy is the most common operation on children in the US.

This is not to say that tonsillectomy is never indicated. But in a majority of cases its necessity is debatable, in particular, since removal of tonsils in children leads to a fourfold higher risk for Hodgkin-lymphoma[20].

Treatment after thrombosis

In the US of the 1950s, patients who had experienced an acute coronary event, a myocardial infarction or a heart attack, were given strict bed rest for about six weeks. Strict really meant strict here. Sitting in a chair was prohibited. Even turning in bed from one side to the other required calling for assistance. Patients were fed by a nurse, so patient care presented a major challenge to the nursing staff.

Medical insistence on rigorous bed rest was based on the transfer of the principle of therapy of a fractured limb to the therapy of a diseased heart. Bed rest resulted in lower heart rate and lower blood pressure. Both together were equated with heart rest and thus with higher likelihood of cure.

Surprisingly, no one had really studied the issue. Due to both this lack of clinical data supporting the present therapy regimen and a gut feeling that staying motionless over weeks could not be beneficial because it would impose severe stress, Bernard Lown, at that time a young physician at Peter Brent Brigham Hospital in Boston, was the first to challenge this strict tenet. He tried to persuade his staff – doctors and nurses – to mobilise the patient on the first day of admission into a comfortable lounge chair rather than lying flat and motionless. However, staff resisted. Most colleagues viewed this as unethical misadventure, because Lown could not present supportive literature, failing to admit that their own claim of correct therapy was likewise unsubstantiated. Over a five month period, against bitter resistance of his colleagues and only by dedicated support of the chief physician, Lown managed to recruit firstly a few and then more and more heart patients to "chair healing".

Compared with patients managed by other physicians with strict bed rest, his patients did remarkably well. On top of that, the workload of the nursing staff decreased substantially. Results were so encouraging that Lown's method was presented at a noteworthy meeting in 1951. Within a few years of its publication, the period of hospitalization was reduced by half and hospital mortality was reduced by about a third. Lown, later one of the most prestigious heart specialists worldwide, stated in 1996[i]: "Considering the fact that in the United States about one million people suffer heart attacks annually, perhaps as many as one hundred thousand lives were salvaged each year by this simple

---

i   This and many more examples of pre-occupied acting in orthodox medicine can be found in Bernard Lowns wonderful book "The lost art of healing" which I would recommend as a duty lecture in medical school.

strategy. ... I continue to be troubled by the ways in which doctors rationalize treatments that are not only without merit but draconian punishments to boot. ... Quietude in the face of a misguided approach relates either to matters of ideology or to an economic advantage." [21].

Cholesterol lowering drugs

In rural areas of China and Japan smoking is not uncommon. In the West smoking often is associated with a higher risk for coronary artery disease (CAD). Yet stroke and CAD are relatively infrequent there. Blood concentrations of so-called "bad" cholesterol LDL (low density lipoprotein) are low on average, most likely due to healthy nutrition and plenty physical working.

High levels of LDL are correlated with higher risk of cardiovascular diseases. To lower LDL, drugs called statins are prescribed. Any adult with LDL above 130 mg/dl (milligram per decilitre) is recommended to take a drug such as Lipitor or Crestor prophylactically. Every day. Success, with respect to LDL, usually is quick: LDL levels fall into what is seen as norm.

But how large, exactly, is the protection against death that cholesterol lowering drugs can provide ? In a large study from 2008 with 17802 overweight patients, half of them got a statin, the control group got a placebo (again, no participant knew what was in the pill). Over 5 years 198 death's (denoted "event" in the study) were observed in the statin group, 247 in the control group. 49 less "events" for statins. In other words, if you are overweight and take a statin, your risk to die within 5 years is 2.2%, without statin 2.8%. Whether this slight risk reduction applies to non-overweight patients remains unclear, but since overweight people have a higher mortality in general, we can presume that the same study done in normal-weight people would result in an even smaller advantage than 0.6%, if at all.

On the other hand, newly diagnosed diabetes occurred in 270 statin patients, 216 in the control group. 54 more diabetics for the group treated with statins. So, as an overweight patient, even if you decrease your risk to die within 5 years by 0.6%, you increase your risk for diabetes by the same 0.6%.

The study was financed by the pharma company Astra Zeneca[22]. Data collection and selection - but not final analysis - was done by the funding company as well. It was not mentioned - which was not the purpose of the study - , that everyone can reduce his risk of death by a healthy life style. Benefit can be more pronounced compared to swallowing a pill.

How many medical doctors, who recommend statins to their patients, know and communicate that a nutrition rich in vegetables and fruits and poor in carbohydrates and salt, combined with regular exercise, lowers the risk of an "event" at least as successful as the drug, at the same time not raising the risk of diabetes above average but lowering it below average ?

Rectal cancer

To treat rectal cancer with radiation after chemotherapy was and is established practice in western countries for several decades. It was believed that additional radiation prolongs survival. This conviction was encouraged by conclusions of the National Institutes of Health Consensus Development Conference in 1990.

However, until 2000, the assertion of beneficial surplus radiation had never been convincingly confirmed. In 2000, a group of scientists set out to scrutinize this assumption for the first time. They concentrated on one particular variant of rectal cancer, namely carcinoma of the rectum stage Dukes-C and Dukes-D. We can assume that the conclusions drawn from this study apply to other cancers of the rectum and the colon as well. 694 patients were enrolled in two groups of about equal size. They were either assigned to postoperative chemotherapy alone or chemotherapy plus post-operative radiotherapy. The result, in the words of the authors, was surprising: "Postoperative radiotherapy resulted in no beneficial effect on disease free survival or overall survival." [23].

Pancreatic cancer resection

Surgical resection of pancreatic cancer has been the treatment of choice for more than 70 years. Pancreatic cancer is one of the worst forms of cancer with very poor prognosis. In one report from 1996[24], out of 684 patients with cancer of the pancreas 118 were resected. Only 7 survived more than 5.4 years. Furthermore the operation is dangerous. According to another study from 1996, 9% of the resected patients died within 30 days post surgery. Thus, one is tempted to ask whether surgery of pancreatic cancer should be recommended when statistical benefit is barely detectable, or in other words, when the balance between potential benefit and harm is so delicate. In a systematic review from 1996[25] on clinical studies published so far, the author concludes bluntly: "In pancreatic cancer 5-year survivors are rare, cure is exceptional, the operative mortality is significant, and the costs of resection are excessive." .

Mammography

A particular delicate and hotly disputed topic is mammography, the X-ray examination of the female breast for cancer screening.

The 2002 WHO/IARC report on breast cancer screening notes: "...the vast majority of women undergoing screening do not have breast cancer at the time of examination, and these women cannot derive a direct health benefit from screening; they only can be harmed".

Non-specialists may understand that 'harm' addresses the X-ray burden here, which might be negligible compared to the potential benefit, namely to detect and treat cancer early. But the WHO-report identifies harm as the problem of false-positive diagnosis, that is the detection of apparent malignant tissue which factually is not malignant, and its implications, namely "the worry associated with a cancer diagnosis and the complications of [unnecessary and harmful] therapy". This is not a rare gaffe: 15-25% of all women with positive mammography are over-diagnosed. In these patients detected nodes are either harmless or not existing or would not lead to fatalities.

Now, how is the relation between potential benefit and harm in mammography ?

Several investigations between 2002 and 2012[26-29] had a similar outcome. The most recent publication[29] showed that 26-30 out of 2500 mammographies lead to diagnosis breast cancer. 10-year survival for breast cancer is about 73%, projected 20-year survival rate about 64%. Of 7 women which would die from breast cancer without mammography, 1 (one) could be saved by mammography (15%). At the same time 6-10 women in the mammography group are treated for breast cancer unnecessarily. If we leave expense and angst aside, unnecessary treatment can only be justified if it does not harm or if benefit outweighs harm. Does it? Most of these women are treated with an aggressive therapy that can induce cancer later.

The problem is that mammography cannot distinguish between fatal and non-fatal nodes, between malignant and benign lesions. A vigorously debated publication from 2008 showed that about 20% of all lesions detected by mammography vanish without any treatment[30]. Possibly this percentage could be increased by fever therapy. We will collect arguments for this claim later.

Thus, the decision pro or contra mammography is not trivial. But at least, one assumes, this discordant situation should be clearly communicated to the women interested in screening.

That is not the case. In another investigation, authors of 143 articles on mammography were classified into three groups: those authors, which perform mammography themselves; those affiliated with screening, for instance working in radiology or surgery specialties; and authors working in specialties unrelated to mammography screening. Within the first group 40 percent downplayed the risk of mis-diagnosis in mammography, 17 percent in the second group and 7 percent in the third group. Hence, the more the authors were involved in screening, the more alleged benefits were emphasized and harms not communicated[31].

Lymph node excision

Surgical removal of malignant breast tumour is often accompanied by removal of some or all of the axillary lymph nodes. If a biopsy reveals malignant cells within the lymph node most proximal to the breast - the so-called sentinel lymph node - axillary dissection including removal of at least ten lymph nodes is obligatory and common practise since hundred years. But these guidelines were developed without any hard evidence, based on so-called plausibility criteria alone. A 2011 study showed that these criteria are incorrect. Removal of axillary lymph nodes does not lead to improved survival and does not lower the risk of relapse[32]. But many patients who have their axillary lymph nodes removed report chronic pain and lymph oedema of the arm.

Prostate cancer

PSA is a protein produced in the prostate and excreted into the ejaculate. Traces of PSA can be found in blood. Measuring PSA concentration in blood is used as laboratory test on prostate cancer. A PSA concentration above 4 ng/ml (nanogram per milliliter) is taken as indication of possible malignant disease.

Other causes can elevate the PSA level in blood as well, for instance a benign enlargement of the prostate or a slowly growing carcinoma which usually does not make a problem during lifetime, or other natural reasons for PSA level fluctuations. Hence the first positive test is usually followed by a second lab test.

If the second test remains positive, usually a biopsy is recommended. If the biopsy reveals malignant cells, urologists ring red alert and recommend quick hormone treatment, radiation or surgery. The latter in many cases leads to erectile dysfunction.

MEDICINE: PLAUSIBLE VERSUS FACTUAL

A large majority of urologists is convinced that early diagnosis enables early treatment, and early treatment raises chances for cure. But is this plausible assumption correct ?

In March 2009 a large study was published in the prestigious journal "New England Journal of Medicine", which did not support this assumption. It ran over more than 15 years and involved more than 76000 males.

This group was divided into two halves of about 38000 men. Men in the first group were offered PSA tests (6 years) and rectal examinations (4 years) at regular intervals, the second group was not examined by urologists. These investigations lasted from 1993-2001. During these and a couple of more years the number of patients dying from prostate cancer was monitored[i]. If the screening and the resulting treatments would convey positive effects, the number of men killed by prostate cancer should be lower in the screened group compared to the control group. However, that was not the case: in the PSA-group 50 men died from prostate cancer, in the control group 44 men. Hence, the screened group fared even a little worse, though not statistically significant[33].

Also, we have to face large over-treatment, that is, therapy recommendation for patients who would never face any problem within their normal life span, let alone die from prostate cancer. It was estimated, that about half of men with diagnosis belong to those without risk[34]. The problem is the same as in breast cancer. We can not distinguish aggressive lesions from those which never lead to any fatality.

Which is going to change. The expense to sequence an entire human genome, which was about 100 Mio $ in 2001, is estimated to drop below 1000 $ in 2013. In a few years we will be able to distinguish dangerous from harmless nodes by sequencing biopsies at low cost. The availability of thousands of genomes will enable a radically new medicine[35].

Back to the present. Testing PSA is enormous expensive. Diagnosis amounts to about 3 billion dollars per year in the US alone. Over-treatment adds more billions. Richard Ablin, who invented the PSA test in 1970, wrote 40 years later, in New York Times March 9th, 2010, in an article entitled "The Great Prostate Mistake": "The medical community must confront reality and stop the inappropriate use of PSA screening. Doing so would save billions of dollars and rescue millions of men from unnecessary, debilitating treatments."

---

i   The error rate for cause of death diagnosis is estimated to be about 1-2%

Yet, it needs courage to tell his patients "whether you do PSA testing is six of one and half a dozen of the other", if the urologist positions himself outside of the majority of his less well informed colleagues, if he questions decade old practise, if he can make good money by recommending the test and when alleged plausibility tells otherwise.

"The human body has no means against cancer"

In his founding talk for the institute in 1965, Professor Bauer, one of the founders of the German Cancer Research Institute (DKFZ) in Heidelberg, claimed that 'the human body has no cancer fighting capabilities'. He wrote, ignorant and arrogant at the same time: „Cancer is the only disease, where no spontaneous healing exists. Cancer allows only a single cure, *sanatio curativa medici*, by the hand of a medical doctor". A contemporary, Prof.Werner Zabel, wrote as a reply: „This statement daunted me. A physician eliminates his best helper, if he does not consider the defence mechanisms of the human body. What surgery is capable of today are admirable technical accomplishments. Respecting these, we should not forget the even greater performance of the organism, which can close wounds and is able to heal".

Bauer was surgeon and an ascetic man. It is said that he slept only few hours per night and did not touch the back of his chair, but always sat stiff and straight. Surgery is the king of disciplines within medicine, surgeons are the primadonnas in the clinic. They get great amounts of admiration not only from patients but also from their peers. If you are a hero, why bother about reading scientific journals ?

Of course the crude claim that many surgeons are stubborn illiterate ignorants would be unjust and a preoccupation on its own, but it might contain a tad truth. Bauer made his comments about humans body weakness at a time when hundreds of reports on spontaneous remissions had been published. There is even some likelihood that during his long career he came across a case of spontaneous remission himself, directly or by talking to colleagues. One year before his inaugural speech in 1964, an article was published in "The Lancet" - one of the prestigious medical journals - describing that people with a certain immune deficiency (ataxia telangiectasia) develop cancer more frequently than average[36]. The highly respected Professor could have known better, if he had wished, his bias was not suited to gain scientific insight, to say the least. And he was not alone. We have to admit that over a major portion of the twentieth century a large fraction of clinical oncologists would simply deny any healing capacity of the human body with respect to cancer.

Evidence based medicine

The American "Institute of Medicine", a representative group of medical experts, estimates that less than half of medical practise is evidence based[35]. Which does not mean the other half is plain wrong. It means, we should be very careful to judge by plausibility arguments alone.

Examples of wrong plausibility as those described above are very rare in the hard sciences mathematics, physics, chemistry, biology[i] but common in medicine. Why ?

The scientific validation cycle - formulating a hypothesis and verifying this by experiments - is much longer in medicine compared to biology and chemistry, where experiments are done on a timescale of days and weeks rather than years as in medicine. Progress is bound to be slow, when new therapies have to be tested in large cohorts of patients over many years.

In addition, clinical trials are immensely expensive and usually only funded by large pharmaceutical companies when good profit is anticipated, that is, when a novel drug is protected by patent. Fever therapy is, from patenting perspective, not "novel technical art", since it was published more than a hundred years ago.

Preoccupation, low profit estimates and a lack of understanding come together in the case of fever therapy. At least the latter, its scientific foundation, has improved tremendously in recent years.

---

[i] To publish evidence of regulatory B-cells in a respected journal with good impact factor - these cells meant a change of paradigm in immunology - was hindered by anonymous reviewers over several years until 2008.

Immunology

Tumour antigens

Many years after the neglected observations of Blacks, Oplers and Speers[17] that the survival rate of gastric cancers correlated with the number of immune cells observed in and around them, it was shown that tumours are invaded and surrounded by a particular class of immune cells, which derived their name "tumour-infiltrating-lymphocytes" (TILs) from this finding. This was confirmed by numerous experiments. For instance, immune cells taken from rats carrying a tumour were labelled with a fluorescent dye and injected back into the rats. The dye was used to pinpoint those cells later under the microscope using a fluorescent lamp. One day after injection, some of these cells were found in liver, kidneys and blood, but a prominent fraction of the labelled immune cells was found within the tumour tissue[37]. This means that the immune system can detect and identify malignant cells, which in turn requires particular signals on the surface of tumour cells[38]. Several of those signals have been identified and are called tumour antigens or T-antigens.

The immune system is capable of localising malignant cells, just as it is capable to localize a bacterium, a virus, a worm or a malaria plasmodium. Still, millions of people die from cancer each year. What happens with bacteria and viruses after detection by the immune system - eradication of the pathogen (in most cases) - does not seem to happen with malignant cells. Step one, detection, works, step two, successful battle, does not work.

Let's call the combat stage of the body-war on cancer "second gear", to distinguish from mere recognition of cancer cells which one might call "first gear". We know that first gear often, if not always, takes place, while second gear at a pace comparable to the fight against a bacterial infection is usually absent. Why does the immune machinery not accelerate into second gear in cancer ? This is one of the outstanding puzzles in biology.

Malignant cells are, after all, derived from the host. They are, as immunologists say, derived from "self", as they arose from normal body cells. Bacteria and viruses come from the outer world, when looked from body's perspective, they are "non-self". The immune- system, during its maturation in childhood, learns how to avoid attacking "self", that is, it learns to avoid an auto-immune reaction. During this learning process a plethora of different T-cells is developed at random. All those, which can detect host-tissue, which are "self-reactive", become eliminated. At the end of the maturation process of the T-cell-system only those T-cells remain, which tolerate "self", that is the tissue and organs of the own body. The remaining repertoire of T-cells is still diverse enough to recognize all sorts of intruders like bacteria and viruses, but the prior selection process ensures that self-tissue is not attacked[i]. This so-called SNS-model (self-non-self) was the textbook model of immune rejection for a long time.

Although cancer cells carry many – in later stages hundreds – of mutations, which distinguish them from healthy body cells, they seem to have a cell surface which still signals "I am not a stranger" or "I am not dangerous" to the immune system. Although it has been shown that many of them also display unusual signposts on their surface – some of them are T-antigens and represent the signposts to attract T-cells – these signals appear not to be enough to alert the immune system fully.

Keep in mind that severe barriers must exist preventing the immune system from attacking its own tissue, otherwise the danger of auto-immune disease would increase. Thus, during evolution, the immune system had to learn to maintain a delicate balance between recognition and removal of pathogens on one side and not attacking "self" on the other side. Cancer cells seem to present a challenge in the middle between these extremes which the immune system is not fully qualified to cope with.

At first glance.

This was the interpretation under the SNS-model. It remains puzzling that often red alert is switched on, as the infiltration of tumours with immune cells indicates.

Experiments with mice

Scientists have bred many different mouse strains to study tumour biology in the laboratory. Like humans, dogs, frogs, sharks, flies and all other living species mice have a genome containing several thousand genes. Nowadays it is possible to grow mice defective in one or a few of these genes, so-called knock-out or KO-mice, to study how important this gene might be for the individual. Using this method, several genes were knocked out by

---

[i] This distinction is, however, not always properly maintained, for instance in auto-immune diseases.

different research groups, genes, which are known to have important functions within the immune system.

Apparently, the immune system is a highly redundant system. In many cases of defects one component can be substituted by other components without any apparent loss of viability

One of the genes tested for knock-out is called perforin. Perforin is active in two families of highly specialized immune cells, so-called natural killer cells (NK-cells) and cytotoxic T-cells (CTL). If the perforin gene is knocked out, these mice are no longer able to produce the Perforin-protein and their immune system is therefore defective at one point. Perforin turned out to be not substitutable. Perforin deficient mice spontaneously developed numerous lymphomas, cancers of the lymph nodes[21]. Later other components of the immune system were identified which upon knock-out resulted in a similar effect: higher incidence of spontaneous tumours .

That a suppressed or damaged immune system leads to more cancer does not look quite amazing, but it is. Turned around, it implies that the normal, healthy immune system is capable of suppressing the development of spontaneous tumours, at least to a certain degree or for quite a while[i]. This has a profound implication: there must exist a second gear, because developing tumour cells are detected and destroyed by a normal immune system. Accordingly, the decade long perception of clinical oncologists, the human body has no cancer fighting capabilities, is clearly wrong.

Nevertheless cancer is a common disease in humans. Despite the proof of principle – the immune system can kill tumour cells – in a majority of cases apparently it does not or not entirely. To be more precise, some tumour cells must be able to escape this defence, and once a tumour has reached a certain size, it would seem that the immune system is overstretched.

Indeed, experiments with mice challenged by transplanted tumours were of mixed success. In the so-called mouse model, where tumour cells are grafted, that is transplanted, underneath the skin of a laboratory mouse to study the immune response, tumour rejection by the immune system can be induced by some extent[ii] within a small time window of some days after transplantation. This usually works only as long as the tumour is not vascu-

---

i   The alternative explanation, perforin-KO-mice are more vulnerable to virus-induced cancers, could be rejected by showing that the incidence of typical viral cancers like lung-adenocarcinoma was not increased.

larized, that is, has not managed to organize for itself a good blood and nutrition supply by inducing its proximate tissue to build up additional blood vessels[39]. This is not quite the same situation as in a patient with clinically evident disease, where the disease developed over month or years.

Tumour infiltrating lymphocytes

The observation described so far and many others suggest a role of the immune system in cancer defence. After organ transplantation and subsequent suppression of the immune system by drugs to prevent transplant rejection, fourfold increases in the incidence of de novo malignant melanoma have been reported. The increased risk for older patients was even 10-fold[23].

Still, overall, the interest in potential beneficial effects of the immune system was low until the 1990ies. Meanwhile, the old observation of Black, Opler and Speer that high numbers of TIL's (lymphocytes around and within tumour tissue) correlate with survival rate, has been confirmed in more than 3400 patients with cancer of the breast, bladder, colon, prostate, ovary, rectum, brain. In the case of breast cancer, the difference was striking. While patients with high numbers of TIL's had a six year survival rate of more than 60%, those with low numbers of TIL's had a six year survival rate of zero percent[40].

Watch out: this is a profound observation again. It is true that the name "tumour infiltrating lymphocyte" is a purely descriptive term. Under the microscope, these cells could be observed in and around tumour tissue, but initially there was no clue about their function[i]. However, the strong correlation between their appearance and survival implies protection, if not more. It implies that even in late stage tumours there must exist an active second gear, a battle and killing of tumour cells to some extent. Longer survival in patients with higher number of TIL's could not be explained otherwise. And the occurrence of spontaneous remission and cure teaches us, that in few cases the immune system can win even in a late stage cancer.

One prominent model derived from the observations described above and others is that the immune system can keep malignant tissue in check and

---

ii   For instance, by injection of known T-antigens (Rosenberg Nature Medicine 2004;10(9):909-15)

i   Only recently it has been shown that at least some TIL's belong to CTL's, i.e. cytotoxic (=killing) lymphocytes (Trojan Lung cancer 2004;44(2):143-7) balance over quite some time, months or even years. During this time constant elimination of some malignant tissue takes place, though not complete eradication. At the same time tumour cells evolve due to their inherent genetic instability. They produce variants leading to successive cell populations with different immunogenicity, that is vulnerability with respect to the immune system. Thus, while one variant cell is detected and destroyed, another variant cell develops, for which the immune system has to generate novel bullets. At the end, in most cases, the outcome is fatal.

How to augment a full blown immune response in later stages of tumour development ? This is the big question. We do not have an answer yet.

Steven Rosenberg, renowned cancer specialist in the US, speculated in 2004 that it might be advisable to "create an inflammatory environment at the tumour site."[39]. Although he does not elaborate how an "inflammatory environment" could be created, he is aware that bringing the immune system into second gear requires processes that were studied in inflammatory diseases. Yet, in his own experiments using cytokines he used to suppress fever, which often accompanies cytokine injections.

Dendritic cells

One important component of a complete immune response are so-called dendritic cells (DC). These cells capture antigens and PRRL-substances, two classes of danger molecules produced by pathogens, the signposts which T-cells need for detection and killing of foreign cells. Antigens are usually proteins, PRRL can be pieces from bacterial cell walls or from viral genetic information. DC „display" antigens and PRRL to the T-cells. To remain in the picture, those signposts initially lie flat and cannot be seen by the T-cells, while dendritic cells serve as street workers to erect the signposts such that they are visible to the T-cells.

It is not obvious why such an intermediate step is required, but it probably has to do with prevention of auto-immune disease. Dendritic cells introduce a second level of control to fine-tune the immune response. They are a mandatory part of the second gear.

Here is a key to understand a full blown immune reaction against cancer cells. Dendritic cells need *two* sorts of danger signals at the same time, PRRL plus antigens, to become activated. Cancer cells produce antigens but no PRRL. Antigens alone - this has been tried without success for two decades - are not enough to alert DC fully.

IMMUNOLOGY

The innate immune system

About hundred years after the experiments of Busch and Coley, we begin to understand what a bacterial infection does to cancer cells. In recent years our immunological understanding has sharpened, both in the area of infectious diseases and in cancer immunology, and we can now build a hypothesis to explain why injected bacterial substances have been such powerful weapons in some cancer patients.

The human immune system can be broadly divided into two parts, the innate and the adaptive subsystem (picture 9). These parts are not physically separated like motor and gear, but consist of different types of immune cells and signals. Both types of cells can be found in all body locations. They have different duties but communicate vividly.

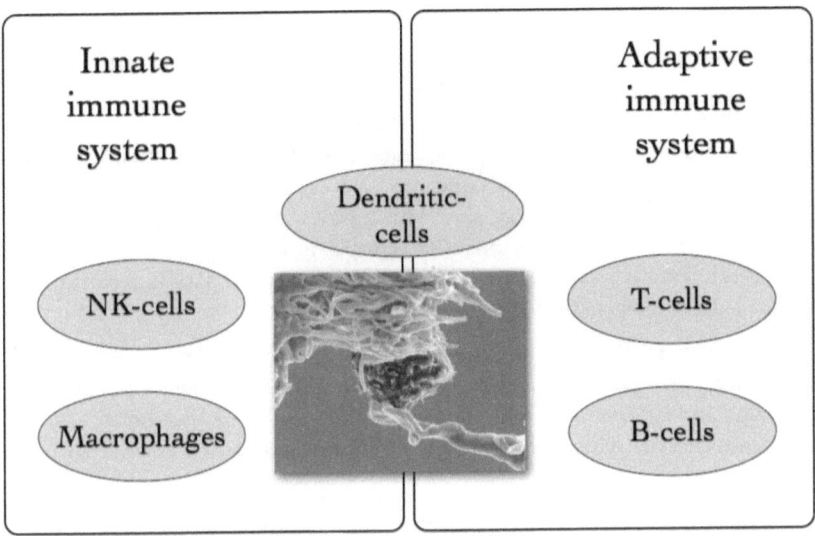

Picture 9: Innate immune system (left): important parts are macrophages and natural killer cells (NK-cells) and the complement system in the blood (not shown). The innate immune system exists in all species. Pathogens cause it to mount a first line of defence within minutes.

Adaptive immune system (right): important parts are T-cells (maturing in the thymus) and B-cells (maturing in bone). The adaptive immune system
- exists in vertebrates only
- requires 4-6 days to be in full gear
- mounts a highly specific immunes defence (e.g. T-cells specific for influenza virus are non-functional against herpes virus and vice versa)
- can produce T-cells which kill infected host cells
- can activate antibody producing B-cells
- can produce memory cells which provide immunity for years.

Dendritic cells mediate between innate and adaptive immune system. These cells patrol in the blood and epidermal tissue and pump tissue fluid in and out to recognize pathogens immediately. Here is the catch: to become activated, dendritic cells require two sorts of danger signals at the same time, PRRL plus antigens. Cancer cells deliver antigens but no PRRL. Antigens alone are not sufficient to activate DC.

IMMUNOLOGY

The focus in cancer immunotherapy for many years was on the adaptive system alone. The adaptive immune system counts as the evolutionary younger high-tech system. It is capable to battle pathogens by targeted attack using antibodies and particularly manufactured T-cells. It can generate immunological memory: memory T-cells travelling for years around the body to fight the same infection faster and harder the second time. To take advantage of the adaptive system, cancer patients are vaccinated with tumour antigens or antibodies. In a laborious variant involving the innate system partly, dendritic cells are gathered from cancer patients, loaded with cancer antigens in the lab and re-injected. The idea is to empower the adaptive system.

These therapies face the same problems. Cancer cells from one patient, even within the same neoplasm, can carry different mutations and are genetically and immunologically distinct. Some cells are bound to escape targeted approaches which address only one or few antigens. Second, PRRL are not considered.

Only vertebrates have an adaptive immune system. Many other organisms - plants, fungi, insects, jellyfish - only possess an innate immune system and are still well capable to deal with pathogens in their environment. One should not underestimate its power and take advantage of it in any therapy.

Meanwhile we know, the innate system is as important as the adaptive, since both work hand in hand. In many cases the innate system is enough to resist an infection and the adaptive system does not get involved. Pathogens entering the body via mucous membranes of mouth, nose, stomach, gut or via little wounds usually are removed by innate system alone. The innate system can respond within minutes. If a wound is larger, an infection can develop which usually damps down without involving the adaptive system either. If the innate system is overstretched, it sends signals to the adaptive system which than requires four to six days to run in full gear and produce antibodies or targeted T-cells. Both weapons are tailor-made and can eliminate one type of pathogen only. Antibodies made against tuberculosis bacteria do not bind streptococci, T-cells against HIV can not help with herpes virus infections.

But even if the adaptive system is finally needed, the innate system is required to slow the spread of pathogens. Mice lacking the innate system are killed in short time by otherwise harmless pathogens which can proliferate explosively.

Yet, many scientists regard the innate system as less important. This lack of attention was the same in vaccinology. Common vaccines such as those to protect from measles, chicken pocks, tuberculosis or whooping cough either contain living attenuated (non-pathogenic) germs, killed germs or pathogen antigens. These substances are meant to stimulate the adaptive system. They lead to the development of T-cells and antibody producing B-cells. Both classes of cells are able to mount an immune response targeted specifically against the invader.

Yet all vaccines contain another component, so-called adjuvants. How adjuvants work was an unknown over decades of successful application. Adjuvants amplify the immune response considerably, some vaccines were almost impotent without these helpers. Charles Janeway, the famous immunologist, mischievously called adjuvants "doctors dirty little secret".

Today we know that adjuvants stimulate the innate immune system.

Cancer immunology, as a discipline, is not as old as vaccinology. But we have to confess the same tunnel view: focus on the adaptive system, neglect for the innate system.

Since discovery of the first tumour antigens in the early 1990ies, in hundreds of clinical trials, it was tested whether by application of tumour antigens an immune reaction against tumour cells could be triggered or enhanced. Other than in vaccinology adjuvants did not play a role in those trials. Overall, the focus on tumour antigen alone was a complete failure.

To this day only a single tumour antigen drug has been approved: Provenge (Dendreon) against prostate cancer. The procedure is technically demanding. Dendritic cells are gathered from the patient by a process called leucapheresis. Cells are sent to Dendreon, where they become incubated by a tumour antigen named PAP, present in 95% of malignant prostate cells, and the immune stimulating protein GM-CSF. Prepared cells are sent back to the hospital and applied to the patient. Success was unboastful. The most successful test achieved an average survival of 26 month, 4 month more than control[i]. Other tests could not confiem any significant advantage. Shrinkage of the primary tu-

---

i   The control group was treated with placebo rather than GM-CSF, which would have been the correct procedure.

mour was not observed. The treatment is immensely expensive (100000$ per patient) and has strong adverse effects. The foundation of Dendreon burned millions of venture capital. Dendreon got the FDA approval in 2010 and declared smashup in 2014.

We could have known better.

Both in mice and humans the innate system is as important as the adaptive, and the same is likely true for all vertebrates. Without the protection of he innate system, pathogens would proliferate much faster during early stages of infection, such that the adaptive system, which requires some days to run full speed, might step in too late. Evolution has wired both parts tightly, they work together dashingly and effectively.

Only recently, cancer immunology is beginning to change focus. One of the molecular canditates for linking spontaneous remissions, epidemiology and fever are PRRL-substances.

PRRL

The innate system is particularly alert at the periphery, in the epidermis below skin, in the mucosa of mouth, nose, ear and gut. Parts of the innate system are very sensible sensors for PRRL-substances produced by bacteria and viruses.

PRRL[i] is a collective term for a range of chemically different substances with a single commonality. They are produced by pathogens only, not by the host, and they signal red immune alert[ii].

PRRL are substances from the cell wall of bacteria like lipopolysaccharide (LPS) or polyglycan, substances from bacterial propellers like flagellin, double stranded RNA only found in viruses or particular substances from infectious fungi like mannan or zymosan. These substances bind to the same

---

[i] PRRL: pattern recognition receptor ligand

[ii] Why PRRL produced by commensal bacteria in the gut do not alert the immune system is an open question. After all, bacteria amount to more than one pound of weight in each human !

protein family in the human body as adjuvants in vaccines: so-called PR-receptors (PRR, thus PRR-ligands are abbreviated PRRL).

PRRL are powerful activators of dendritic cells. In fact, there is no other class of substances known which can induce maturation of dendritic cells as efficiently as PRRL. Hence dendritic cells are mainly activated upon bacterial or viral attack, while cancer cells are no good activators of dendritic cells.

Let's recollect. The innate immune system responds quickly. If pathogens cannot be eradicated completely, the adaptive system jumps in after a few days. Cross-talk between both systems is mediated by dendritic cells, which in turn are alarmed by PRRL-substances plus (!) antigens, both stemming from bacteria, viruses or fungi.

But even if we understand that PRRL signal the presence of pathogens and activate dendritic cells, which are required for a full immune reaction against cancer cells, where is the link between PRRL and cancer ? What about the timely connection between feverish infection and spontaneous regression ? In cancer we have cancer antigens, during infection we have bacterial or viral antigens, and in the second case one would expect that defence is built up only against pathogen.

Not quite. If PRRL are sensed by dendritic cells, they collect antigens indiscriminately, whether it is tumour antigen or antigen from pathogen, and activate all sorts of T-cells. The same activation of multiple T-cell clones happens with multivalent vaccines (for instance MMR, a childhood vaccine immunizing against measles, mumps and rubella in a single shot).

Dendritic cells travel using the blood stream and through lymph nodes, but they also squeeze through dense tissue. They are the never pausing guardians of the immune system. On their way patrolling tissue they permanently pump tissue fluid in and out. In the proximity of tumours, tumour antigens are swallowed. Antigens alone cannot activate dendritic cells full gear. Only if antigens and PRRL are taken up simultaneously, dendritic cells become activated, walk into the lymphatic system and "present" antigens to T-cells (picture 10), in order to activate these executors. In patients with an infection presumably both cancer antigens and pathogen antigens are presented, each to different T-cells. PRRL from one sort of antigens (from pathogen) presumably can serve as activation signal for two types of T-cells (against cancer cells and pathogen). Without PRRL, both sorts of antigen could not lead to any T-cell response. Thus, PRRL allow for a sort of cross-reactivity[i] which might offer unleveraged potential in cancer therapy.

Drew Pardoll, cancer immunologist at the John Hopkins University in Baltimore, Maryland, and his co-workers found an interesting detail in 2004[41]. Often, established cancers find ways to tune down the immune response in their environment and later in the whole body, by producing immune suppressing signals and releasing them into their environment, a phenomenon called 'tumour escape' or 'tumour tolerance induction'. Pardoll was interested to 'break' this tolerance, that is to revitalize a normal immune response against an established tumour (they worked with mice). His group administered dendritic cells plus tumour antigen, but tolerance against the antigen remained. In a second experiment the dendritic cells were infected with a virus, and now tolerance against the cancer antigen was 'broken', the immune system launched a full attack against the antigen. Please note that the virus helped to launch a full attack against a cancer antigen. Of course, viruses produce PRRL, so the picture painted above is supported by this experiment: dendritic cells are fully activated only by help of PRRL. The observation that many cases of spontaneous regression had a connection in time and perhaps in cause with a fairly severe feverish infection naturally leads to the speculation that dendritic cells might in general be involved in these puzzling cases of spontaneous healing.

Infections can trigger tumour rejection. This was taken advantage of in the experiments of Busch and Coley and others more than hundred years ago, although the explanation of Coley, Streptococci might contain some magic anti-cancerous substance, was wrong.

It also makes sense from another point of view. Dendritic cells capture their signposts, which are needed to activate T-cells, from both infected and dying cells. Heat stressed cells can die, depending on the level of heat. Cancer cells are more vulnerable to heat than normal cells, many will die[42]. Fever produces heat, so it is fair to assume that PRRL plus fever produces an unusual

---

[i] Bakteria can stimulate macrophages to deliver co-stimulatory signals. These signals can activate T-cell with non-bacterial antigen specifity (cross-presentation) (Nature Reviews 2002; 2:185-194)

high amount of cell debris from cancer cells. Fever, most likely, generates signposts.

If this were true in cases of established tumours – as the occurrence of spontaneous regression implies – and since cancer is usually a slowly progressing disease often with long periods of dormancy, putative beneficial fever effects should also precipitate as preventive efficacy. This is indeed the case.

But before we leave immunology and have a short walk in epidemiology, let's have a brief look at mistletoe therapy. Recently it has been shown one of the main actors in mistletoe extract, mistletoe lection, is a PRRL.

IMMUNOLOGY

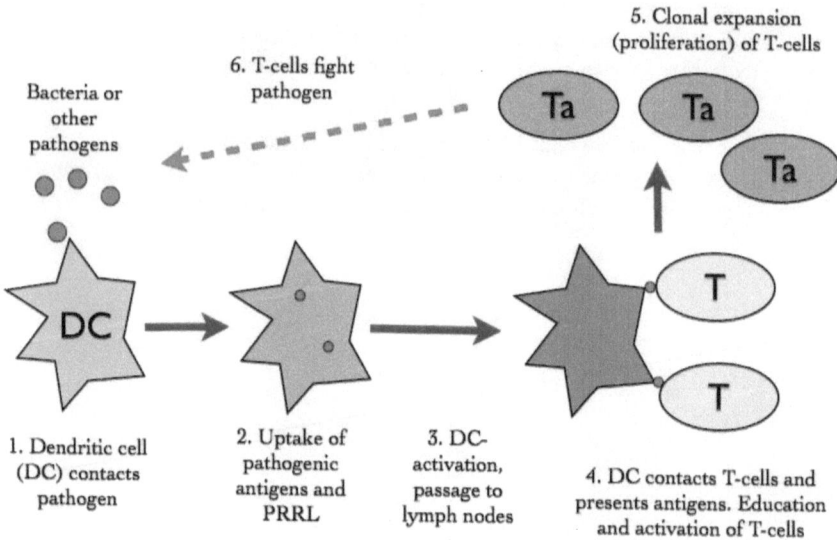

Picture 10: Two-step activation of T-cells. T-cells can recognize cancer antigens and cancer cells, but they have to become activated by dendritic cells. Dendritic cells, in turn, require antigens plus PRRL-substances for their own activation.

Mistletoe therapy

Many cancer patients in Europe ask for mistletoe adjuvant therapy. According to Schwabe Arzneiverordnungsreport 2010 (Springer, Heidelberg) in Germany alone mistletoe was applied about 23000 times per day.

The mistletoe plant was introduced to the cancer field by the founder of anthroposophy, Rudolf Steiner. Mistletoe is a parasitic plant growing on trees. Steiner regarded mistletoe as a sort of plant ulcer, not unrelated to a neoplasm, and by applying mistletoe extract to cancer patients he meant to treat "equal by equal". This naive conclusion, by luck or intuition, resulted in a true drug.

The main actor in mistletoe extract is mistletoe lectin, a toxic and immunogenic protein. Some publications indicate that the effects of mistletoe extract cannot be fully resembled by the lectin, which would mean that additional immune stimulating compounds are in the extract like viscotoxins, which are similar to snake toxins, or bacterial exotoxin impurities. The brand Iscador, which can be applied i.v. and s.c., contains endotoxins due to its particular manufactoring procedure.

Proteins are chains of amino acids. The order of amino acids - 20 different can be used - in a particular protein is called its sequence. The sequence of mistletoe lectin is known, it contains 266 amino acids. This sequence can be compared to the sequence of millions of other proteins stored in biological databases.

Mistletoe lectin (ML) is remarkable immunogenic. Usually, only pathogens - bacteria and viruses - produce such immunogenic substances, not plants. To understand how the strong immune stimulatory activity of ML could be explained, I started a sequence comparison in the database.

The protein most related to mistletoe lectin is ricin. Ricin is produced by the castor bean, the source of ricin oil, and is very toxic (the oil is detoxified by heating). One quarter of a milligram ricin can kill an adult. In fact, ricin is so toxic that it is on the list of substances prohibited in biological warfare.

Both proteins, mistletoe lectin and ricin, are composed of two parts, the A-domain and the B-domain. The A-domain determines toxicity and is almost identical in both proteins. The B-domain determines, to which sorts of cells the protein can bind and thereby whether the A-domain can be transported into the cell. Only inside can the A-domain can do its toxic work. B-domains of mistletoe lectin and ricin are different. Ricin is not more dangerous than mistletoe lectin because it is more toxic but because it can be bound and ingested by more cell types.

These database comparisons usually result in a long list of more or less related proteins. Far down in the list, at the border between faint sequence similarity and random hits, I found a protein which brought me up short. Shigatoxin.

Shigatoxins are proteins produced by *Shigella dysenteriae* and EHEC-bacteriae. Shigella can cause bad dysentery, EHEC both dysentery and severe haemolysis. In spring 2011 53 people died in north Germany from an aggressive EHEC variant.

Proteins are not stretched chains of amino acids but folded into sophisticated 3D-structures. It is the structure of a protein which determines its function, in this case its toxicity. The sequences of mistletoe lectin and shigatoxin are very different. Usually, such diverse sequences end up in different 3D-structures and different function. The situation is similar to two sentences of different words but identical length, which are written on top of each other, letter by letter. Even if some letters can be found at the same place - for instance an "A" at position 29 , a "H" at position 53 of both sentences or whatever -- the meaning of both sentences will be different.

Despite the sequence difference of both proteins, I looked into another database - the 3D-database of protein structures - whether mistletoe lectin and shiga toxin were listed. Indeed, both structures have been resolved and stored into the database a couple of years ago. I loaded both structures into a program to compare them. To my great surprise I found that the structures of both A-domains are identical. The A-chains of Shiga-toxin and mistletoe lectin have precisely the same shape. This is similar to two sentences of different letters and having the same meaning. A very rare case in lyrics, a very rare case among protein structures.

How can it be that a bacterial toxin is found among the genes of a plant ?

The phenomenon is called "horizontal gene transfer". Long ago in the evolutionary past, probably millions of years ago, mistletoe (and ricin) have captured a bacterial gene and incorporated it into their own genome. Perhaps to rebuff animals to eat the plant. In the course of evolution, the amino acid sequences diverged. But the function of the respective proteins, shiga toxin and mistletoe lectin, obviously were so important, that evolution kept the structures unchanged.

If mistletoe produces a bacterial toxin, the immune stimulatory activity of mistletoe lectin is immediately plausible[43].

Early test tube experiments indicated a dosage optimum with respect to immunogenicity, such that high doses resulted in immunosuppression[44]. But the notification about possible immune suppression has been removed from the German state institute for drugs and medicinal products (Bundesinstitut für Arzneimittel und Medizinprodukte, BFARM) list of possible mistletoe side effects, since such effects have never been confirmed in humans.

In 2010 a Korean group found that lectin from the Korean mistletoe is a PRRL substance[45]. European mistletoe lectin, which is used in drugs such as Iscador, Helixor and several others, is highly similar to the Korean variant, so it is a PRRL as well.

Mistletoe lectin is a non-self protein. It is not produced by the human body. It should be an antigen. Indeed most patients under mistletoe therapy develop lectin antibodies after few applications. In principle, these antibodies should neutralize the lectin. But immune-stimulatory effects persist despite antibody production. Patients report positive effects after months and even years. In one case an inoperable tumour which was monitored by ultra sound over years always started to grow if mistletoe therapy was interrupted over longer intervals and halted growth under mistletoe therapy. Many aspects of mistletoe effects are not well understood.

Several clinical studies on mistletoe therapy, which induces mild fever, were published[46]. A majority of studies found extended survival, most studies reported clear improvements in quality of life. However: there is no systematic investigation yet on the benefits in therapy-naive patients. We know that standard therapies - chemotherapy, radiotherapy (but not surgery alone) - compromise the immune system not only during therapy but up to two years after cessation[47,48]. Most patients ask for mistletoe after standard therapy has been finished. So mistletoe can unfold its immune stimulation only in part until the aftermath of chemo and radiation petered out.

It would be possible to calculate the full potential of mistletoe therapy if those thousands of patient data lingering in dark subsurface clinic rooms were evaluated by focussing on the small cohort of patients which, for whatever reason, had no prior chemo. This has not been done.

IMMUNOLOGY

After introducing mistletoe and showing that its main ingredient, mistletoe lectin, is a pathogenic substance originating from bacteriae, let us now turn to the prophylactic effects of feverish infections.

Cancer epidemiology

A hefty feverish infection and spontaneous remissions obviously go together in a majority of cases. If this correlation were causal rather than accidental, it should also precipitate as prophylactic affect. People with more feverish infections should have a lower incidence of cancer. Within the huge avalanche of cancer literature it is indeed possible to find publications indicating that infections lower the risk to develop cancer later in life (see table 1) This conclusion does not hold for chronic infections, which can even induce cancer. It is mandatory to distinguish chronical and acute infections: while chronic infections tone the immune system down, acute infections activate the immune system.

Observation	Effect	Year of publication	Pathogen	Ref.
Lower risk of cancer in syphilitic prostitutes	prophylactic	1725	Treponema pallidum	49
Collection of 302 cases of spontaneous regression (44 complete remissions) 27/302 cases accompanied by infection (9%), 69 cases where "incomplete operation [was] often accompanied by post-operative fever" (28%)	therapeutic	1918	diverse	13
Low risk of cancer in tuberculosis patients	prophylactic	1929	Mycobacterium tuberculosis	50
Lower risk of cancer in malaria patients	prophylactic	1929	Plasmodium falc., malariae, vivax	51,52
Of 300 cancer patients 113 had no febrile infectious childhood diseases (FICD), while in 300 controls 16 lacked FICD	prophylactic	1934	diverse	53
Fewer childhood diseases, higher cancer risk in adults	prophylactic	1936	diverse	54
In a cohort of 300 cases of childhood leukaemia, 26 spontaneous remissions were observed. 21/26 (80%) were accompanied by infection	therapeutic	1951	diverse	15

CANCER EPIDEMIOLOGY

Description	Type	Year	Organism	Ref
"...according to the Cancer Centre in Sao Paulo (Brazil), among tens of thousands of cancer patients only two gave a positive Machado reaction [indicating chronic or recovered Trypanosoma infection], whereas among the remaining population the number suffering from this infection varies from 10 to 20 percent.", anecdotal remark	prophylactic	1963	Trypanosoma cruzi	11
Lower cancer mortality in 5460 survivors of typhoid fever	prophylactic	1970	Salmonella typhi	Ref.58 in 55
Fewer physician visits, secondary illnesses and hospital referrals in 150 controls vs. 150 cancer patients	prophylactic	1970	diverse	56
In 62/224 cases of spontaneous regression (28%) either an infection or a persistent temperature elevation was observed prior to regression	therapeutic	1971	diverse	16
Occasional remissions in Hodgkin's lymphoma after measles attack	therapeutic	1971	Morbilliviruses	57
Patients developing empyema after lung cancer surgery have improved 5-year survival (50% (n=18) vs. 22% (n=411))	therapeutic	1972	diverse	58
Smaller incidence of mumps, measles, rubella in 300 patients with cancer of the ovary compared to control group	prophylactic	1977	diverse	59
Increased cancer risk with an odds ratio of 2.6 for missing history of infectious organ diseases, 5.7 for missing history of common colds and 15.1 for missing history of fever	prophylactic	1983	diverse	60
Out of 353 individuals with a negative history of measles 21 developed cancer versus 1 case in 230 controls with a positive history of measles (p 0.001)	prophylactic	1985	diverse	61
Much lower cancer rate in wool and hemp factories; wool or hemp dust can carry bacterial endotoxins.	prophylactic	1985	diverse	62
Lower frequency of infections in the first year of life for children with leukaemia	prophylactic	1986	diverse	63
Lower cancer incidence after Herpes infections	prophylactic	1987	Herpes simplex	64
Post-transfusional hepatitis in patients with acute myelogeneous leukaemia doubles survival rate	therapeutic	1982, 1992	Hepatitis viruses	65,66

A history of common colds or gastroenteric influenza was found to be associated with a decreased cancer risk (odds ratio 0.18 and 0.23 vs. population and hospital controls, resp.)	prophylactic	1991	Common cold viruses	67
Inverse correlation between number of infections and mortality from tumours in Italy in the period 1890-1960: each 2% reduction in number of infectious diseases was followed by a 2% increase in tumours about 10 years later	prophylactic	1998	diverse	68
Inverse association between number of carcinoma (but not breast cancer) and febrile infectious childhood diseases (FICD); association stronger for higher numbers of FICD and childhood in pre-antibiotic times; strongest protection by rubella (379 cancer cases vs. 379 office matched controls)	prophylactic	1998	diverse	69
68 well documented cases of spontaneous regression from melanoma, preceded in 21 (31%) cases by a febrile infection	therapeutic	1998	Streptococcus pyogenes	70
Statistically significant inverse association between a reported history of infections and glioma, meningioma (RR=0.72, age and gender matched population control of 1509 cases)	prophylactic	1999	diverse	71
Inverse correlation between melanoma risk and number of recorded infections on one hand and between melanoma risk and fever height on the other hand, leading to a combined reduction of melanoma risk of about 40% for people with a history of three or more infections with high fever above 38.5°C (age and gender matched population control)	prophylactic	1999	diverse	72
More than two-fold higher incidence of cancer in Europe, GUS and US compared to Africa and Asia of 381 vs. 156 (ten most prominent cancer forms, age standardized rate per 100000 population; in Africa and Asia a significant higher rate of infections is assumed here	prophylactic	2003	diverse	73
Prior immunisation of melanoma patients with vaccinia or BCG is associated with better survival (age matched controls)	prophylactic	2005	Vaccinia, BCG vaccine	74

CANCER EPIDEMIOLOGY

Dairy farmers, but not crop and orchard farmers, report one third less cancers than the average population; protection diminishes over time after exposure is removed; dust in cattle houses can carry bacterial endotoxins which frequently lead to unspecific "day fever"	prophylactic	2005	diverse	75
The 10-year survival for patients with osteosarcoma with infection within one year after surgery (n=41) was 84.5% compared to 62.3% in the non-infected group (n=371)	therapeutic	2007	diverse	76
After allogeneic stem cell transplantation, patients who had a febrile infection (FI) before post-transplant day 21 (FI group) had a lower probability of leukemic relapse (P < 0.001) and a higher relapse-free survival rate (P = 0.012) than those patients who did not have a FI before post-transplant day 21 (non-FI group)	therapeutic	2008	diverse	77
4-fold higher risk for Hodgkin-lymphoma if tonsils are removed at age < 15 years	prophylactic	2010	diverse	20
Reduced ALL (acute lymphoblastic leukaemia) risk in kindergarten children (frequent mutual infectious contaminations presumed), OR 0.8) or children with repeated common infections (OR 0.7)	prophylactic	2010	diverse	78
Reduced risk for ALL in children visiting kindergarten	prophylactic	2010, 2011	diverse	79, 80
Reduced HL risk (Hodgkin lymphoma, 128 cases aged 5-14) and NHL (non-Hodgkin lymphoma, 164 cases aged 2-15 years) vs. 1312 controls. HL+kindergarten: OR 0.5; HL+common infections+non-breast-feeding: OR 0.3; NHL+birth order 3: OR 0.7; NHL+prolonged breast feeding: OR 0.5; NHL+frequent farm visits in early life: OR 0.5; NHL+asthma: OR 0.6	prophylactic	2011	diverse	81

Table 1: Anti-correlation between acute, cured infections and the likelihood to develop cancer. Two publications were found which could not confirm inverse association between infection and cancer [82,83], one, in a low impact journal not listed in PubMed [84], reported an increased risk with mumps and whooping cough. All three publications are based on less than 200 cases.

Many clinical oncologist would confirm this statement: "My patients were never sick", meaning that a majority of cancer patients over years or even decades before cancer diagnosis were free of any infectious or feverish disease. Yet, in most studies cited in table 1 fever is not mentioned or only incidentally. A possible auxiliary role of fever is not within perspective. The connection between spontaneous regression, fever, cancer epidemiology together with an immunological explanation linking all these was published only few years ago [85].

The inverse correlation between cancer and infection is only a correlation in time. A causal connection is possible but not proven. But together with our knowledge about the immunological effects of fever, about the innate immunes system and the epidemiological data one can formulate an umbrella hypothesis stating that infections or better feverish infections can delay the onset of cancer - possibly by eliminating pre-cancerous cells - , and in the case of a diagnosed malignant disease infections can extend survival (see "therapeutic effect" in table above) and even exert curative effects (in rare cases of spontaneous healings). A hypothesis involving PRRL as mediators of all these effects has been published 2008[86]. If this hypothesis is correct, one should be able to utilize fever and PRRL in cancer therapy.

Cancer is the most extensively explored disease in the world. Is it possible, that the link between acute infections and cancer has been overlooked ?

Yes and no.

No, because one can find many voices in the medical literature pointing into the same direction.

Parmenides (about 540-480 BC) said: "Give me the power to induce fever, and I cure any disease".

Apparently, in the middle of the 19th century medical doctors treated blood cancer patients by fever induction, which were continued over longer periods[87]. The outcome of these experiments have been forgotten - this question might be an interesting diploma thesis for a medical historian.

Julius Wagner von Jauregg treated syphilis patients in the 1920ies by infecting them with malaria and inducing recurrent fever. Different malaria pathogens lead to forms of malaria with different severity. The dangerous, even life threatening and most common form of malaria is caused by *Plasmodium falciparum*. Malaria tertiana (fever every three days) or malaria quartana (fever every four days) are relative benign forms, caused by *Plasmodium vivax*, *Plasmodium ovale* or *Plasmodium malariae*. Von Jauregg used the latter, weaker pathogens. As a result, the survival rate of syphilitic patients jumped from one to thirty percent. Von Jauregg received the Nobel price in medicine in the year

1927 for his work. His method was applied until antibiotics were implemented in the late 1940ies.

Shear, in 1950, observed remissions in about 10% of children with untreated leukaemia, about three out of four remissions occurred after an acute infection. In a remarkable lucid statement he wrote: "Are pathogenic and non-pathogenic microorganisms [perhaps] one of Nature's controls of microscopic foci of malignant tissue, and, in making progress in the control of infectious diseases, are we not removing one of Nature's controls of cancer ?" [88].

A pioneer in unconventional cancer therapy was the physician Josef Issels, who in Germany in the 1950ies applied some seemingly strange treatments, which from todays perspective were remarkably perceptive. He supposed, cancer is the result of a weak defence status of the body, thus his therapy program was aimed at strengthening this defence. His "whole body approach" included a diet containing raw fruit and vegetable, proteolytic enzymes, vitamines and minerals; fever induction using bacterial extracts; hyperthermia; ozone therapy; re-application of patients own blood; but also low dose chemotherapy. He insisted to remove dead tooth and infected tonsils, since he observed that his therapy worked much better after removal of foci of possible chronic infection. This aspect is noteworthy and should be investigated as well as another suggestion. He observed that he could reduce the dosage of chemotherapeutical drugs by half without loosing efficiency, if the drug was applied during the temperature maximum of a hyperthermia treatment. His repertoire included psychotherapy to remove emotional obstacles and daily jogging. Today we know that moderate physical activity relaxes the mind and stimulates the immune system.

Issels suggested to treat cancer along dual tracks. The "symptomatic" therapy should deal with the local tumour tissue, while the "basic" therapy should re-vitalize the immune system.

According to his own records he treated about 370 cancer patients with a five year survival rate of 87 percent (this number cannot be verified any more).

The medical community isolated and blocked Issels. He was accused of cheat and man slaughter and sent to prison in 1960, but was rehabilitated completely later.

Issels was not a scientist. His explanations on the origin of cancer were wrong, his biological and immunological knowledge limited. But obviously he was a healer, not only physician, a man of strong intuition. Today we know that many of the methods he combined make immunological sense.

Some of his observations and conclusions were way ahead of his time. He was right in criticising that "the training of defence mechanisms of the body is withheld from people by rash application of antibiotics and analgesics to suppress fever", an insight obtaining acceptance only very recently, after decades of shooting antibiotic-bullets on little disease sparrows. "It is understandable that physicians want to be members of a scientific discipline. But one has to acknowledge that medicine also always has to be guided by experience. [...] The view of medical school trained oncologists, cancer is healed once the tumour is radically removed, is wrong. [...] Patients are treated by knife, X-ray and chemicals, but almost nothing is done to strengthen their defence potential, neither before nor after conventional local treatment. [...] A considerable amount of the huge expense in cancer treatment should be diverted to investigate the immune-biological circumstances, which make the body capable to dismiss cancer" [50].

Thousands of patients have been treated by Issels methods during the 1970ies and 1980ies in German private clinics by his co-workers. Unfortunately, results were never evaluated or published.

In the 1960ies and 1970ies some academic groups tried to test Coley's method again using commercial bacterial extracts like MBV or Vaccineurin. But in contrast to Coley, mainly patients pre-treated by conventional methods were tested and fever was regarded as "toxic" side effect and suppressed[14,89]. Results were not consistent. There were some remissions, even some cures, but the majority of patients did not benefit. The experiments were stopped.

Results from chemotherapy or X-ray therapy are more obvious: tumours can disappear. That cancer as cause of death did not decrease much during the last fifty years[90] is out of account, or people hope to be not far away from a breakthrough in cancer therapy.

So voices similar to those mentioned above can be found in the literature. However, these voices have infrequently been spoken out, have not been heard by many and were not taken serious by many.

Yes, the co-incidence between a feverish bacterial or viral infection and cancer regression has been largely overlooked.

How fever "works" is not well understood. A causal link between fever and cancer regression seemed unfeasible. The hypothesis, substances produced by pathogens can activate dendritic cells, which in turn can activate tumour specific T-cells is new and not yet arrived in medical circles.

CANCER EPIDEMIOLOGY

In clinical everyday routine fever is considered useless, a nuisance to patients and staff. As we have seen before, pre-occupied medical wisdom can be oppressive. Fever accompanies dangerous infections, so its removal is equated with successful prevention of danger. Its "guilt by association" is deeply entrenched in the minds. Fever patients have to be monitored carefully in the clinic, because a proliferative infection can cause circulatory problems. So why not stop it using an Aspirin or an analgesic suppository, which is applied easy and quick and tolerated without problem.

To claim benefit resulting from infections seems to be challenged from another perspective: chronic inflammation was estimated to contribute to 15–20% of all malignancies[91]. But the data are very clear: we strictly have to distinguish chronic from acute, fully cleared infections. We know that acute infections activate the immune system, while chronic infections tone it down.

Fever induced by sterilized pathogens or pathogenic substances is much less dangerous than a proliferative infection. Circulatory problems caused by Vaccineurin, a Streptococci containing drug phased out in the mid 1980ies, were extremely rare. Fever induced by bacterial extracts usually last only one or half a day and than declines automatically.

PRRL-therapy

Past clinical tests with PRRL substances

Getting approval for Coley's extract would be difficult today, for good reasons. Complex substance mixtures are not easy to standardize, meaning, it is almost impossible to guarantee always the same composition. Living bacteria grown in fermenters could, in principle, all of a sudden produce a substance that causes bad side effects. It is much easier to get approval for single substances produced in chemists lab. Dosage and purity can be controlled. It may turn out, however, that at the end of the day with single substances we cannot reach the effectivity of an extract.

> The main function of vitamin C is to protect against oxidation. Anti-oxidation can be measured easily in the lab. It turns out that the anti-oxidative effect of an apple is way larger than the vitamin C within the apple. The anti-oxidative protection 100 gram of apple provide is 260 times larger than the anti-oxidative effect of the vitamin-C in the apple. In other words, to get the same protection as 100 gram apple containing 5.7 milligram vitamin C, one needs 1.5 gram pure vitamin C. It is not the vitamin C alone that provides protection but the collaborative effect of a plethora of substances in the fruit[92].

Likely its not a single PRRL alone which is responsible for the power of Coley's extract. Remember that he started to see successful treatments only after he added a second pathogen, Serratia, to the Streptococcus extract.

Some PRRL can be synthesized in the lab. Some were tested in the clinic. It was not the old Coley story, which led to testing PRRL clinically. Most clinicians do not even know that there might be a connection between PRRL and Coley. PRRL, in clinical tests, usually are regarded as adjuvants, as little helpers for other drugs, similar to adjuvants in vaccines. If during the test a fever occurs, it is treated as toxic side effect and the fever is suppressed, the dosage reduced or the test stopped. Lectures to be learned from the old Coley experiments, from the connection in time between feverish infections and spontaneous regressions, from epidemiological findings, those lectures are not taken.

PRRL-THERAPY

PRRL, without exception, are tested in patients after or during standard therapy, with compromised immune systems. In one case some positive results were seen in a phase-I study which could not be reproduced with a larger number of patients in phase-II, likely because the patients in the latter study had an even worse immune status[93].

Fever is suppressed in these studies.

During an acute infection the body is challenged with a mix of different PRRL substances. Presumably the body mounts a stronger immune response against a mix of PRRL rather than a single substance. We know from vaccinology that attenuated or live pathogens cause a much stronger immune response than single antigens. In the clinical studies so far single substances were tested.

Like with other cancer drugs in testing, patients are taken out of the study if the disease progresses, when a tumour does not stop growing or sometimes even when it stops growing but does not shrink (only since 2010 the FDA acknowledges growth stop as therapeutic goal). With immunological treatments, often a stabilization of disease can occur which, if it is durable, can be regarded as success[94]. From the beginning a longer treatment with several applications per week should be planned, independent of common "therapeutic end points".

PRRL are applied systemically, that is intravenously. But we have hints that stimulators of the innate immune system can be much more powerful when they are applied where the antigen is, namely close to the tumour.

Meanwhile the textbook SNS-model of immune rejection is challenged by an alternative model, the so-called danger-model [95]. According to this model the innate immune system must be stimulated again and again, probably over weeks and month, because - unlike the adaptive system - it has no "memory". This model is in striking accordance with Coley's practise to stimulate over and over again. In the present studies PRRL are applied only a few times, these studies are still driven by the crude perception of a "magic bullet" drug which cures fast or not at all.

How a fever therapy respecting all lessons under modern standards in therapy naive patients would perform is an unresolved question. It may remain unresolved.

No physician can conceive to dissuade a patient from standard therapy as long as the advantage of an alternative treatment has not been proven. The proof, however, can only be obtained if a group of patients abstains from standard therapy. A circular trap.

Let us, despite this principal problem, try to find therapeutic windows for PRRL therapy.

Therapeutic windows for future clinical tests

To test a new therapy which involves the immune system, ideally patients with intact immune system should be selected. Otherwise the full therapeutic potential may remain hidden. In subsequent tests it may turn out that, against the odds, patients after chemo may also profit. But for the establishment of the therapy an un-compromised immune system will allow best benefit.

For some forms of cancer it is uncertain whether chemotherapy or radiation lead to more benefit or more damage, because their application is not correlated with longer life expectancy. Pancreatic cancer and colon cancer were mentioned before. After successful surgery of a colon cancer, adjuvant chemotherapy may be recommended or not. That depends on the responsible physician[i]. To design a clinical study with two groups - one group treated by chemotherapy and one by PRRL-therapy - probably would not be unethical (informed and signed patient consent is obligatory).

Even less problematic, from an ethical standpoint, are early neoplasms which have been completely removed, for instance black melanoma. Usual aftercare consists of patient skin monitoring in regular intervals. Since melanoma are very immunogenic forms of cancer, a comparison between observatory monitoring and PRRL adjuvant therapy seems obvious.

Another therapeutic windows might be prostate cancers, which usually grow slowly. In elder patients prostate cancers statistically do not decrease life expectancy. Prostate cancer can easily be monitored by ultrasound and lab markers, such that a clinical study would compare a control group under observation and a group treated by PRRL therapy.

Glioma, like prostate cancers, of low grade (WHO I and II) usually grow slowly, often over several years, and can remain untreated - there is no effective treatment anyway -, as long as they do not cause symptoms.

---

i   Physicians recommendations, in turn, may depend on the amount of "research money" the clinic gets from drug manufacturers for each patient channeled into chemotherapy

About 5% of non-Hodgkin's-lymphoma are primary cutaneous B-cell lymphoma. Of those, two subgroups (follicle center and marginal zone B-cell lymphoma) tend to be indolent and have a 5-year survival rate of more than 95%[96]. Systemic progression is rare. It has been observed that intra-lesional injections of interferon-alpha or interferon-gamma can lead to remission in non-injected distal lesions[97]. These forms of cancer likely are immunogenic malignancies amenable to other immunological treatments like fever therapy.

Finally, there are forms of cancer where no standard therapy with clear healing potential exists, for instance pancreatic or liver cancer. Pancreatic cancers usually are treated by chemotherapy only briefly, if at all. If no remission can be achieved, physicians know that chemo can be stopped early to obviate severe adverse reactions. Access to a clinical PRRL study could be offered to those patients.

Approved PRRL substances

Drugs applied to humans in a regular clinical study have to be manufactured under GMP-guidelines (GMP: good manufacturing practise). Manufacturing under GMP costs about 200 k€ to start substance production. Combining substances costs a multiple. For drugs applied to lab animals or to humans under Therapiefreiheit (see below) only GLP quality is required (GLP: good laboratory practise). GLP manufacturing costs a fraction of GMP. Many PRRL substances can be ordered in GLP quality.

GMP substances used in on-going clinical studies are not approved and not available in pharmacies, but are usually available for additional clinical studies from drug manufacturers, in particular if novel therapeutic windows can be opened (table 2).

PRRL (receptor)	Manufacturer	GMP / GLP
CRX-527, CRX-675 (TLR-4)	Corixa / GSK	GMP
CpG ODN (TLR-9)	Invivogen	GLP
dSLIM (TLR-9)	Mologen	GMP
E5564 (TLR-4)	Eisai	GMP
Imiquimod (TLR-7/8)	Invivogen	GLP
Loxoribine (TLR-7)	Invivogen	GLP
LPS (TLR-4)	Sigma	GLP
LTA Staph.aureus (TLR-2/6)	Sigma	GLP

MALP-2 (TLR-1/2)	Enzo	GMP
MPL (TLR-4)	Corixa / GSK	GMP
PAM2CSK4 (TLR-2/6)	Invivogen	GLP
poly-AU (TLR-3)	Ipsen-Beaufour	GMP
poly-ICLC (TLR-3)	Hiltonol / Oncovir	GMP
poly-Us21+DOTAP (TLR-7)	Innate Pharma	GMP
Resiquimod (TLR-7/8)	Enzo	GMP
Vaximmune (TLR-9)	Coley-Pharma/Aventis	GMP
Zymosan (Dectin-1, TLR-2)	Invivogen	GLP
Stimuvax (enthält MPL, TLR-4)	Biomira/Merck	GMP

Table 2: PRRL and manufacturers (selection, no liability assumed).

Since substance listed in table 2 are either produced in GLP quality not approved for application in humans, or are reserved for clinical trials supervised by the manufacturer, it is important to remember that several approved drugs available in normal pharmacies contain PRRL and may induce fever, as judged by the respective instruction leaflet (table 3).

Brand	Manu-facturer	Ingredients	Main indication	Fever reported as adverse event	Approved for cancer therapy
BCG	CC-Pharma Medac	Attenuated live *Mycobacterium bovis*	vaccine	Yes	Yes
Bron-cho-vaxom	Eurim Pharm	Lyophilised extract from *Haemophilus influenzae, Diplococcus pneumoniae, Klebsiella pneumoniae* und *ozeanae, Staphylococcus aureus, Streptococcus pyogenes* und *viridans, Neisseria catarrhalis*	respiratory infection	No	No
CADI-05	Immu-vac	Autoclaved *Mycobacterium indicus pranii*	leprosy	Yes	Yes
Cholera vaccine	Wyeth	inactivated cholera bacteriae	vaccine	Yes	No
Colibi-gen inject	Laves	metabolic products from *Escherichia coli laves*	colon inflammations	No	Yes
Detox	Biomira Inc.	adjuvans, contains MPL (monophosphoryl-Lipid-A) anond	adjuvans	Yes	No

PRRL-THERAPY

Name	Company	Description	Indication	Rx	Cancer
		Extrakt aus der Zellwand von *Mycobacterium phlei*			
Ixiaro	Novartis-Behring	inactivated enzephalitis virus	vaccine	Yes	No
JE-VAX	Sanofi-Pasteur	inactivated enzephalitis virus	vaccine	Yes	No
MPL	Corixa	MPL from *Salmonella minnesota*	adjuvans	No	No
Picibanil	Chugai	lyophilised *Streptococcus pyogenes*	cancer	Yes	Yes
Pollinex	Ben-card	pollen allergenes and MPL	allergies	No	No
Polyvaccinum forte	IBSS biomed (Poland)	inactivated extract from *Staphylococcus aureus, Staphylococcus epidermidis, Streptococcus salivarius, Streptococcus pneumoniae, Streptococcus pyogenes, Escherichia coli, Klebsiella pneumoniae, Haemophilus influenzae, Corynebacterium pseudodiphtheriticum, Moraxella catarrhalis*	chronical and recidiv. inflamm. process. of the respiratory tract, bladder and endometr.	Yes	No
Pyrogenalum	Medgamal (Russia)	LPS from *Salmonella typhi*	nerve trauma, prostatitis, uretritis, uveitis, latent TBC	Yes	No
StroVac	Strathmann	inactivated *Escherichia coli, Morganella morganii, Proteus mirabilis, Klebsiella pneumoniae, Enterococcus faecalis*	Recid. bladder inflamm.	Yes	No
Typhoral	Novartis-Behring	*Salmonella typhi*; apathogene Lebendkeime und inaktivierte Keime	vaccine	Yes	No
YF-VAX	Sanofi-Pasteur	yelow fever virus	vaccine	Yes	No
Zylexis	Pfizer	inactivated *Parapoxvirus ovis*	veterinary drug, immune stimul.	No	No

Table 3: Approved PRRL-drugs (selection, no liability assumed)

For Polyvaccinum and Pyrogenalum in particular the occurence of fever of up to eight hours duration is documented as common adverse effect in the respective instruction leaflet. These approved drugs cost only few Euros in their home countries Poland and Russia and might be first choice as bacterial extract replacement and for combination with mistletoe.

Practical application of a PRRL-mix

A physician in Germany may, under a couple of provisions, decide which therapy she chooses for the benefit of the patient. Similar rules apply in other countries. Provisions for freedom of therapy ("Heilversuch under Therapiefreiheit") are due diligence, consideration of state-of-the-art therapy and patient's consent after elaborate instruction. Heilversuch can be claimed for instance if no approved therapeutic option with adequate chance of cure exists. In this case the physician may take the responsibility to choose a plausible therapy. Plausibility can be admitted when cases with positive outcome have been described or if the treatment makes scientific sense. Financing by health insurance must be negotiated for each individual case, or the patient may decide to pay on a private basis.

To simulate a proliferative infection, more than one PRR - receptors for PRRL on immune cells - should be addressed. A PRRL-mix should be used. Since mistletoe-lectin is a PRRL[45], a combination of mistletoe extract with other PRRL suggests itself. In particular physicians with experience in mistletoe therapy acknowledge fever - in contrast to many peers -, and know how to handle it. For instance, Mistletoe, Colibiogen and BCG are approved drugs in cancer therapy. They contain PRRL and can induce fever. Why not try to combine them?

Suppose, we want to apply a PRRL-mix or a drug combination such as mistletoe, Colibiogen, Polyvaccinum. How should we dose and apply the substances? The following hints are a gist derived from older case studies (Busch, Coley, Klyuyeva and others), from reports made available to me by biotech companies and private clinics using self-made Coley-extracts or mistletoe extract, guidelines from a 2008 meeting on fever therapy, and immunological aspects[86].

Exclusion criteria

According to the guidelines for hyperthermia and fever therapy resulting from the 4th meeting "Aktive Fiebertherapie" 2008 in Baden-Baden, Germany, fever therapy is not indicated in cases of

PRRL-THERAPY

- Acute infection
- Heart or circulation insufficiency stadium III or higher
- Acute condition after lung embolism, infarct, thrombosis
- Ulcus ventriculi sive duodeni
- Epilepsy
- Acute psychosis
- Anaphylaxis
- [added] drug addiction
- Patient selection

At higher risks for adverse effects or greater strain during fever therapy and thus potential exclusion criteria are

- hypotension (systolic < 90 or diastolic < 65)
- impaired kidney or liver function, indicated for instance by creatinine clearance
- bad thyroid parameters or medication for hyper- or hypothyroidism
- impaired lung function (asthma, chronic respiratory disease)
- a history of cardiovascular disease (check EKG)
- underweight
- a very weak general condition
- CD4 count below 200
- laboratory blood counts, albumin, alkaline phosphatase, bilirubin, bicarbonate, calcium, sodium, LDH, phosphorus, potassium, total protein, CRP outside norm

In these cases fever therapy should be considered only if the respective physician has ample experience.

In the Hufeland-Klinik in Germany, which applies fever therapy since decades, common practise is to prepare weak patients for fever therapy by initial whole-body hyperthermia.

Smoking should be reduced or stopped during fever therapy. During pregnancy a higher risk of spontaneous abortion cannot be excluded. Breast feeding should be discontinued during fever therapy, since nutritional milk composition might change.

Fever lowering drugs like aspirin or paracetamol and immune dampening drugs like cortisone or opiates certainly should be avoided during PRRL treatment.

If the patient anticipates tumour surgery, it may be advantageous to start PRRL-therapy before surgery, when tumour antigen levels are high and a strong immune activation can be anticipated.

Therapy duration

From the beginning a longer therapy should be planned, lasting over several weeks. At the beginning an indiuvidual fever inducing dosage should be determined. After initialization, ideally three to five injections per week for 1-2 weeks and two to three injections per week for another 2-3 month - a "metronomic therapy". The experiments of Coley and Klyuyeva have shown it, and it makes immunological sense: the innate immune system, lacking memory, must be stimulated again and again by a sham proliferative infection. T-cells activated by dendritic cells up-regulate TLR (PR-receptors), i.e. become capable to "read" PRRL signals on their own. This TLR expression is transient and gradually downregulated over the course of some days without further PRRL signalling[98]. Since a proliferative infection provides a constant level of PRRL, it seems obvious to resemble such a constant supply by PRRL therapy and keep dendritic cells and T-cells in an activated state.

Individual dosage determination

Fever therapy might be best started before surgery, when the level of tumour antigen is high, and continued afterwards. Individual infected patients generate different levels of fever. Some people produce high fever upon weak infections, others produce hardly any change in body temperature even with a heavy lung infection. Accordingly, Coley used to start with a minimal dosage and doubled dosage with each following application. For PRRL substances other than mistletoe extract a starting concentration of 1-10ng/kg BW (nanogram per kilogram bodyweight) presumably should be low enough. The dosages for mistletoe therapy are well established, as are the dosages for the bacterial extracts used in private clinics such as the Hufeland-Klinik and the Tijuana clinic. If the patient developed fever of 39°C-40°C, Coley kept dosage constant in subsequent applications. If the fever became too high - above 41°C - or the patient was too weak, dosage was reduced. If the fever decreased over several applications, dosage was increased again. This was a truly personalized treatment.

> Pharmaceutical companies try to develop and market such personalized treatments under the label "pharmacogenomics", but we did not quite arrive at a true personalized medicine yet.

Coley had the feeling - and case analyses by his daughter Helen Coley-Nauts 50 years later appeared to confirm this feeling - that higher fever correlated with better outcome. However, this was never rigorously confirmed. It might turn out that a moderate body temperature elevation combined with a broader PRRL-mix or the addition of other immune stimulators to the mix or some other variation of the procedure results in similar or even better outcomes, compared to Coley's time. We do not know.

Coley explicitly warned to apply the *first* injection intra-tumoural, but rather to start with s.c., p.t., i.m. or p.t. applications.

- Patients receive only one application per day.
- Blood pressure, heart rate and temperature should be monitored starting before application and for at least 6 subsequent hours, or longer if temperature remains above 38°C after 6 hours.
- Applications should be done two hours after a light meal or on empty stomach.
- Subsequent applications are only permitted if the body temperature has returned to normal (< 37.5°C) and about 24 hours have passed since the last infusion.
- Dose should be increased if previous dose lead to a body temperature elevation of less than 39°C.
- If a dose which lead to high fever before does not achieve high fever any more, dosage should be increased again.
- If a dose which lead to fever above 39°C before, subsequently leads to higher fever above 40.5°C, dosage should be reduced for the next application.
- For very weak patients a fever target range of 38°C-39°C might be preferable to therapy truncation.
- Appropriate dosage adjustment should continue for the entire course of fever therapy.

If the body had been prepared for PRRL i.v. applications, tumour lesions, which are accessible for injection, might be injected peri-tumourally, that is intra-muscularly a few centimeter distant from the tumour edge. I.v. and p.t. injections can be combined. The starting dose for p.t. injections should be low again and increased according to the scheme for initial applications. The location of p.t. injections can be gradually moved towards the lesion and finally continued as i.t. injections. I.v. injections can and should be gradually abandoned if p.t. or i.t. injections alone result in body temperature elevations above the target range (39°C).

Occasionally, a patient may experience difficulties in igniting a high body temperature. For these patients an external source of warmth, like a hot bath, hot water bottles, placement in a warm room may result in higher fevers at lower doses.

Maintenance phase

After induction phase, patients are to receive 3-5 applications per week for 3-4 weeks and 2-3 injections per week for another 3-4 weeks. A further 3-6 month of reduced intensity treatment may be advisable, depending on outcome. Since the optimal duration of maintenance therapy is individually different (depending on tumour load and prior therapy, for instance), these recommendations are only educated guesses.

If the patient has a treatment pause of 7 days or more, dosage should be reduced 2-4fold or more for the next treatment.

Maintenance therapy can be continued in ambulant or outpatient setting, if body reactions have been found to be steady and robust.

Auxiliary measures

Patients should be packed into a blanket and feel warm before infusion. Together with a hot-water-bottle this measure limits the energy expenditure required to maintain a high body temperature. Uneasy side effects such as headache, chills and nausea are avoided or reduced. Recurrence of those side effects are a signal to reduce dosage or avoid food and fluid intake two hours before injection. Vitamin-D3, calcium and magnesium supplements before infusion can help reduce side effects.

- Heavy meals should be avoided during fever, drinking water in small portions should be encouraged.

- Fever above 41°C can be alleviated by calf packing and, in case of longer duration, with aspirin.
- Appetite in the evening when fever is down is counted as a good sign and should be pleased.
- Alcohol should be avoided or limited to one drink per day and is permitted only after fever is down.
- Patients should be kept warm after fever declined to channel energy towards the now stimulated immune system rather than body temperature maintenance.
- Patients should rest during times of chills, rigors and fevers to reduce the risk of orthostatic hypotension.
- Nausea and headache can be treated homeopathically with Nux vomica D3 and Tabacum D12 for one day and oxygen inhalation.

Side effects

Patients will often experience chills within 1-2 hours after infusion start, with underweight patients at greater risk. Shaking chills typically last no longer than 15-40 minutes and can be alleviated or prevented by auxiliary measures (see above). Using bacterial extracts or LPS[i], fever will usually peak 2-3 hours after initiation of chills; with mistletoe fever onset can be delayed until 18 hours after application.

Before treatment starts, patients should be informed about other possible side effects:

- Sub-cutaneous or intra-muscular application can lead to local skin irritations or signs of inflammation. Severity of skin irritations can be reduced by subsequent massaging the injection site. S.c. or i.m. application has the advantage of a depot-effect leading to slower but longer immune stimulation. Most manufacturers of mistletoe extract recommend s.c.-application.
- If fever exceeds 41°C, anti-pyretic measures and mild calf cooling can be considered. The next dosage should be reduced 2-4fold and / or delayed for 2-3 days.

[i] Lipopolysaccharide 2ng/kg BW i.v. (ref. 111)

- If infusion causes a body temperature decline beyond normal, similar dosage adjustment and / or pause are necessary. Decrease in temperature may be the result of an excessive dose or occur in debilitated patients or those with a weakened immune system.
- Strong adverse effects like nausea, vomiting, diarrhea should result in similar dosage adjustments. If fluid loss is severe, fluid should be replaced i.v. by 0.9% sodium chloride solution.
- In cases of other severe (grade 3 or 4) adverse symptoms fever therapy should be discontinued.
- PRRL therapy can lead to steep increase of tumour cells dying, in particular in tumours with good blood supply. On one hand this can be interpreted as a good sign of immune defence ignition. On the other hand kidneys can be overstretched by the load of cell debris, which can lead to tumour-lysis syndrome. Treatment should be abandoned until kidney lab markers stabilize.
- Pain in tumour lesions can be observed during the chill phase, often followed by a decline of pain below pre-injection levels.
- Sometimes the fevers lead to transient bone pain, which Vitamin-D3 and calcium/magnesium supplements help to reduce.
- During chills cutaneous vasoconstriction and cyanosis may be observed. Again, these can be alleviated by proper warming.
- Fatigue and sleepiness are common and expected.
- Excitability and irritability short after injection resolving after the chill phase are usual.
- Increased heart rate are commonly seen during chills.
- Myalgia, arthralgia and hyperesthesia are common.
- Dry mouth has been reported.
- Generally there tends to be a mild decrease in blood pressure during therapy. Hypertension or hypotension may occur shortly after injection in patients who are not adequately warmed. Faintness is seen after abrupt rising during chills.
- Anorexia, adipsia and weight loss occur often during fever and resolve once fever has declined. Patients often report increased appetite after one week of therapy.
- Photophobia may occasionally be noticed.
- Headache may occur.
- Impaired cognitive functioning is normal during high fevers.
- Menstruation disturbances have been reported.

- Accidental intravenous injection may lead to immediate rigors, shortness of breath, rapid heart rate, hyperventilation and / or nausea. Symptoms can be alleviated using diazepam. Fever may develop normally afterwards.
- Fever usually is followed by marked leucocytosis

Signs of positive outcome

In vascular, ulcerating or fungating tumours rapid degeneration can occur, often signed by sloughs. In less vascular tumours changes observed are softening, reduction in size, fluctuation. Following i.t. injection a transient increase in size, with skin becoming red and tense, may be observed.

In clinical studies treatment is stopped if the tumour continues to grow37. This practise would require reassessment for fever therapy, at least for the first weeks of treatment. It is known that the activated immune reaction can lead to a massive influx of immune cells into the tumour (tumour infiltrating lymphocytes, TIL). More TIL, better prognosis[17,40]. Up to 40% of tumour volume may be immune cells. But this influx can lead to a temporary enlargement which, without biopsy, can hardly be distinguished from malignant growth. Physicians with expertise judge the success of fever therapy by general patient conditions: does pain decrease, do blood markers improve, does the patient report more energy or easier mobility or better appetite. In cases of palpable tumours softening is a sign of improvement.

Fever will be followed by a marked leukocytosis. Cytokine markers which increase following proper stimulation of the innate immune system include $TNF-\alpha$, IL-1, $IL-1\beta$, IL-6, IL-12, $IF-\gamma$. Inflammatory markers and markers of immuno-suppression such as IL-10 and $TGF-\beta$ should decline in the long run. The so-called neutrophil-lymphocyte ration should fall below 4 [110]. These laboratory markers for innate stimulation can be monitored over the first weeks and might indicate beneficial treatment - this has to be investigated.

Risk of severe adverse reactions

It cannot be stressed enough: fevers induced by PRRL-containing drugs do not persist in the same manner as fever caused by a proliferative infection may persist. Body temperature will inevitably decline after 12-24 hours, depending on dosage. Since dosage is optimized on a per-patient basis, starting with very low dosage, the final dosage can be determined, aimed at a decline of body temperature elevation within 12 hours.

Over 30 years and along several thousand applications, Coley reported six fatal cases following fever therapy in his own department and another three fatal cases from colleagues. He presumed that these nine cases were "probably or possibly" caused by the treatment with bacterial extracts. All patients had inoperable late stage tumours. Two i.v.-injected patients died from embolism. Three patients got a too high initial dosage, one of the three directly into the tumour. Three patients died from nephritis, likely caused by tumour lysis syndrome. In one case a second injection was applied during high fever. Hence six out of nine fatal cases could be counted as medical malpractice and avoidable. The three cases of tumour lysis syndrome likely would not be fatal in modern clinical settings.

If all relevant aspects are considered, fever therapy does not harm. Fever is an evolutionary old mechanism, tested and certified since million years.

Yet, fever therapy will be no walk in the park. It requires a dedicated patient and a dedicated physician.

Hyperthermia

Fever induced by PRRL or bacterial extracts can be regarded as an active fever. The body generates heat and the innate immune system becomes stimulated. In contrast, by hyperthermia the body is heated from the outside, with little or no immune activation.

- If the whole body with exception of the head is warmed up to about 40 °C by infrared radiators, the procedure is called whole-body-hyperthermia.
- Local hyperthermia aims at targeting malignant tissue exclusively. Heat is generated by microwaves or ultrasound. Target temperature is about 44oC in order to destroy tumour cells.
- Another strategy is to increase tumour blood flow by external heat, to improve access and efficacy of cytotoxic drugs.

The report of the German „Gemeinsamer Bundesausschuss über die Bewertung der Hyperthermie" 2005[i] was quite sceptical. Whether hyperthermia alone can decrease cancer mortality is unclear. Hyperthermia combined with chemotherapy or radiation leads to longer survival[99] or higher response rates[100], compared to chemo or radiation alone. Presumably hyperthermia can lend a hand to active fever therapy, because it reduces the energy demand of the body to generate heat.

Yet, Hyperthermia could play a role in augmenting active fever therapy. Some patients have difficulties to develop fever or high fever, which might be easier with help of whole body hyperthermia. Also, hyperthermia and hot water bottles can reduce the metabolic expense the body has to invest for producing high temperature, such that active fever therapy is less exhausting.

Since cancer cells are less resistant to higher temperature compared to non-malignant cells, hyperthermia could reduce tumour load and raise tumour antigen levels, such that a PRRL-induced immune reaction could be stronger.

---

i   http://www.g-ba.de/downloads/40-268-236/2005-06-15-BUB-Hyperthermie.pdf

Cancer fever therapy: cases

Fever therapy using bacterial extracts is still done in some private clinics, for instance the Hufeland-Klinik in Bad Mergentheim in Germany, the „Klinik im LEBEN" in Greiz, Germany, the GISUNT-clinic in Wilhelmshaven, Germany, the Lukas-Klinik in Arlesheim near Basel in Switzerland or the Issels-clinic in Tijuana, Mexiko. They use privately manufactured extracts without official approval for cancer treatment. Physicians have to ask for particular permission for each patient (Switzerland) or treat under the principle of therapy freedom ("Therapiefreiheit", Germany).

Mistletoe adjuvant therapy, inducing less strong and shorter fevers, is applied to thousands of patients each year in Europe.

The success rate clearly is lower than in Coley's time, most likely because a majority of patients is heavily pre-treated, or because frequency and duration of treatment are sub-optimal. Health insurances do not pay PRRL-therapy treatments lasting many weeks, so treatment will not be long enough. Often patients take cortisone to reduce pain and local inflammation. Cortisone suppresses the immune system on top of the damage from chemo and radiation.

Still, occasionally, favourable outcomes can be observed.

CANCER FEVER THERAPY: CASES

Case 1 - B-cell lymphoma

MBVax-CF application is not permitted in the US and in Canada but has been applied for compassionate use in a couple of other countries. Between 2006 and 2012, CF was distributed by MBVax to interested physicians in several countries in Europe and Asia. In 2013 MBVax stopped distribution of CF and is recruiting venture capital to build a GMP facility for CF production.

Dosage of CF is determined on a per-patient basis, starting with a very low dose and stepwise twofold increase until fever of 39°C is achieved, as described above. Subsequently, dosage is kept or adjusted if fever decreases.

After analysis of about 80 cases, MBVax claims about 20% full and 70% full and partial remissions.

A patient with a large B-cell lymphoma and several metastatic lesions started MBVax-CF treatment by age 39 in mid 2010. Previously, chemotherapy, radiation and stem cell transplantation had been tried without avail. The patient received intramuscular injections of CF (into pectoral muscles) initially, and all subsequent treatment was CF intravenously. He was treated two times per week for about one year, achieving fevers of 39°C or greater. Remission between April 2010 and July 2011 of a thyroid lesion causing considerable discomfort, of a mediastinal lesion in the upper deep breast and another nasal soft tissue lesion (see picture 11) were documented by PET-computer tomography. He continued monthly maintenance therapy after mid 2011.

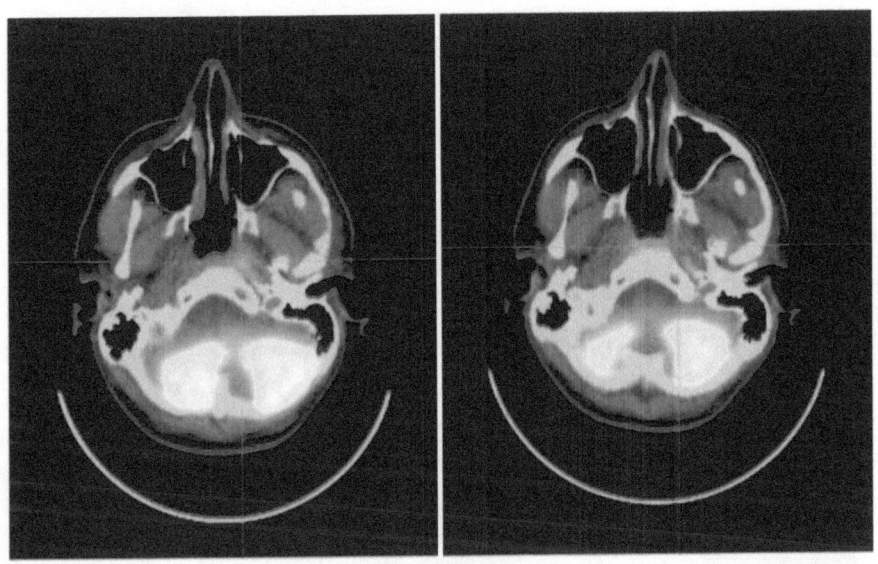

Picture 11 (case 1): Transversal view on nasopharyngeal soft tissue lesion before and 15 month after CF therapy

Case 2 - Non-Hodgkin lymphoma

In 2005, a 42-year old male patient developed swollen lymph nodes in the neck and was diagnosed with follicular lymphoma. He tried various alternative treatments which for a while resulted in stabilizing the disease. In late 2007, however, a large tumour in the kidney led to kidney dysfunction, so-called hydronephrosis. He began MBVax-CF therapy in April 2010. The patient used CF intravenously two times per week over more than a year, achieving fevers of 39oC and higher. The remission of two lesions until September 2011 was documented by computer tomography (see picture 12 for the kidney lesion).

The patient was in good health with minimal residual disease post treatment. He continued maintenance therapy.

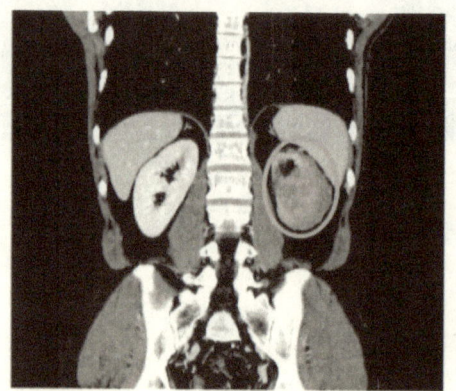 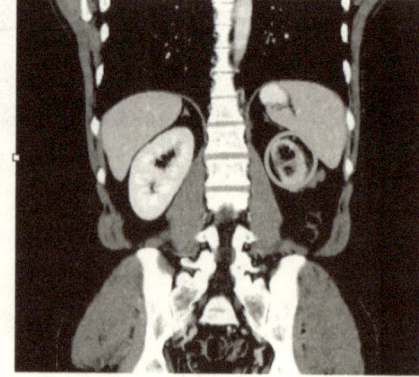

Picture 12 (case 2): Remission of kidney tumour between March 2010 and September 2011

Case 3 - follikular lymphoma

A 62-year old patient with follicular lymphoma, over 17 years was treated using CHOP, Rituxan, Bexxar, RICE and Humax. In 2008 she had one large and several smaller masses in the abdomen, several large nodes in groin and smaller nodes on the clavicle. She was suffering from liver pain, impaired kidney function due to hydronephrosis, itchiness and painful hand and foot cramps when she started MBVax-CF treatment in mid 2008. The patient received intra-tumoural injections of CF into the groin region, generally 2-3 times per week until fall 2011, achieving fevers over 38.5°C. During the treatment, groin lesions regressed. Until November 2011, abdominal tumours cleared (see picture 13) and symptoms resolved completely.

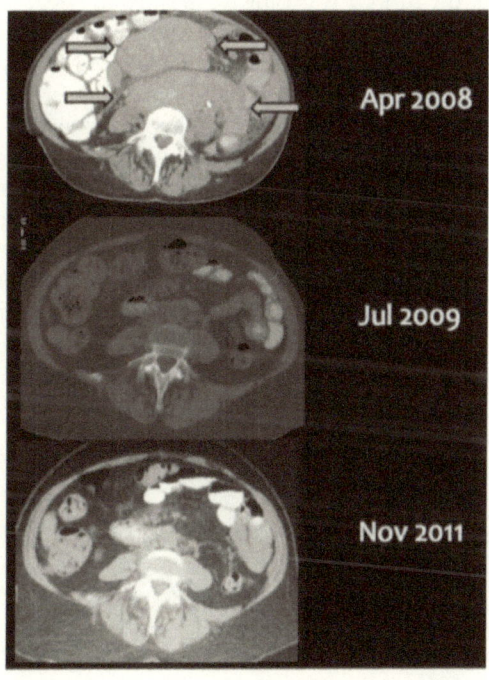

Picture 13 (case 3): Complete regression of massive abdominal tumours between April 2008 and November 2011.

Case 4 - Oesophageal cancer, lung recurrence

In November 2002, patient underwent esophagectomy with gastric pull through for esophageal adenocarcinoma. In early 2007 at age 63 years, recurrences were noted in the left upper lung and pancreatic head. In March 2007, before starting MBVax-CF treatment, a mass associated with pancreas was 7 cm large. Level of carcinoembryonic antigen (CEA), a laboratory marker for adeno-carcinoma in adults, was 224 (normal between 0 and 2.5 micrograms/liter in non-smokers, up to 5 in smokers). He began CF therapy in May 2007. By September 2007, after several months of treatment with CF, the mass haemorrhaged. Patient received 30 units of blood in 30 days. Site specific radiotherapy stopped bleeding. Patient restarted CF in January 2008. By September 2008, the mass was greatly diminished in size, only a small focus of FDG[i] avidity remains. By December 2007 with no treatment, the lung mass regrew from 3.0 cm to 3.8 cm. Patient recommenced CF therapy in Jan 2008. In September 2008, patient's lung mass remained. In November 2008 he received cyberknife therapy[ii] to the lung mass and began CF therapy once more. In May 2009, mass was grown in size to 5.2 cm. In September 2009 size was stable at 5.2 cm. Pyloric dilation adjacent to pancreatic head was noted in September 2009. Biopsy found a poorly differentiated adenocarcinoma. He began 10 weeks of FOLFOX[iii] in Nov 2009, followed by CF therapy up to June 2010. A stent was placed, but the pre-pyloric lesion was no longer visualized. Patient received no further treatment after discontinuing CF therapy in June 2010. Scan in July 2010 shows no FDG avidity in lung lesion. PET/CT showed no evidence of active disease. In June 2012, CEA was 1.8 (normal <3.1). Throughout the period of CF therapy beginning in 2007, patient remained active biking, skiing, and taking long motorcycle trips. He was in excellent health in 2013 (see picture 14).

---

i   FDG: fluorodeoxyglucose, radiopharmaceutical used in PET imaging

ii robotic radiosurgery system designed for more accurate delivery of radiation to tumours compared to systemic radiotherapy

iii chemotherapeutic combination of folinic acid, 5-fluorouracil, oxaliplatin

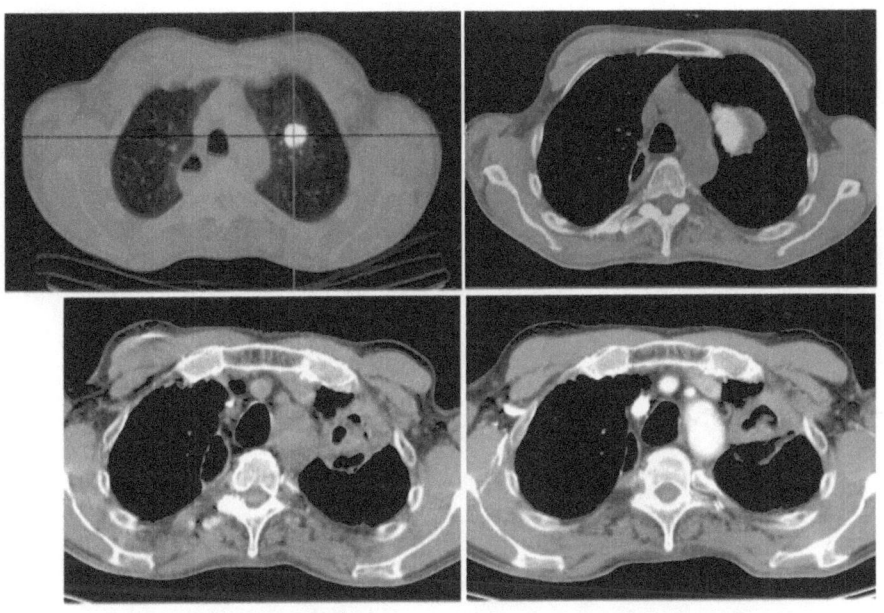

Picture 14 (case 4): Remission of a lung metastasis after combined CF and chemotherapy.
Pictures taken March 2007, September 2008, June 2011, February 2012

Case 5 - Malignant melanoma

A 21 year old female patient developed melanoma in her scalp in 2007. She received wide excision, but there was a recurrence in the right parotid gland. This lesion was excised and followed by interferon (IFN) therapy. A PET-CT in August 2010 showed a recurrence which was again excised, but followed by a new lesion in the post-auricular region. This was followed by right mastoidectomy and a right auriculectomy and radiation to the tumour bed. Recurrences in the head and lungs were followed by MBVax-CF therapy starting February 2011. A July 2011 PET scan revealed no evidence of tumour activity (see picture 15).

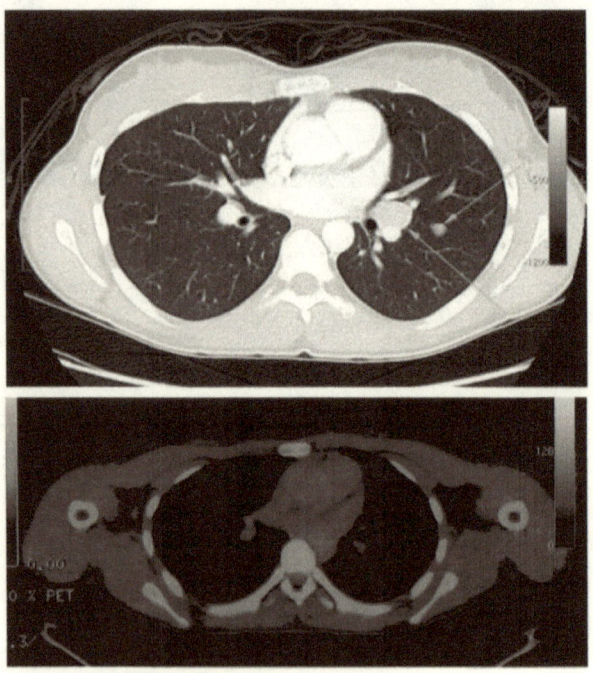

Picture 15 (case 5): Top: PET-CT image of 2.2x1.4 cm metastatic left hilar node and left sided metastatic 0.9x0.6 cm pulmonary nodule in January 2011. Bottom: resolution of left hilar node and pulmonary nodule in July 2011

Case 6 - Breast cancer

A 39 year old female patient was diagnosed in early 2009 with cancer of the left breast in the upper quadrant. The patient underwent left mastectomy with axillary node dissection (28/50 nodes positive). She was treated with 3 cycles of fluorouracil, epirubicin and cyclophosphamide; radiotherapy to the left chest wall, supraclavicular fossa with post axilla boost; and finally 8 weeks of tamoxifen. Subsequently, she was diagnosed with metastatic breast cancer, with lesions noted in the C7 vertebra and right hip and a possible lesion on left third rib. She began regular MBVax-CF-therapy in June 2011. In late 2012 she was in excellent health and continuing on CF therapy (see picture 16).

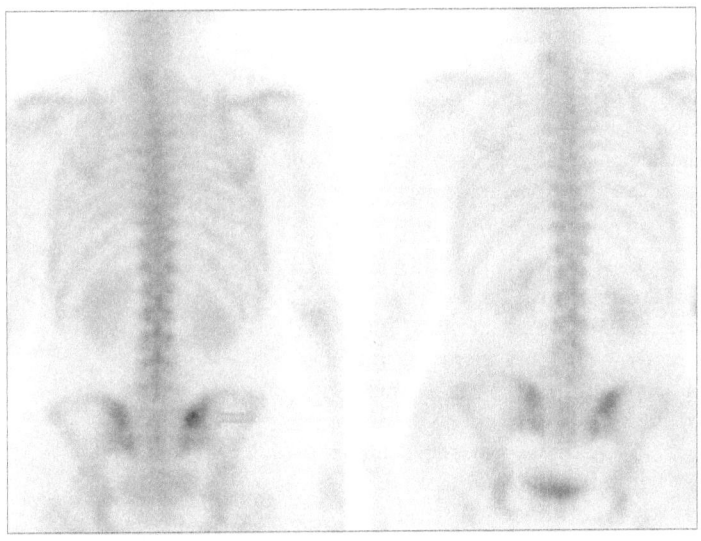

Picture 16 (case 6): shadows in right hip march 2011, undetectable in november 2011

CANCER FEVER THERAPY: CASES

Case 7 - Melanoma lung metastasis

In the Issels-clinic in Tijuana, Mexico, a specialized cancer treatment protocol is applied lasting four weeks which can be continued, in parts, upon discharge. This protocol includes extra-corporeal photopheresis with dendritic cell vaccine, Coley's mixed bacterial vaccine to induce fever and stimulate innate immunities and metabolism (this preparation is manufactured by a local facility), Issels autologous vaccine, auto-hemotherapy, photoluminescence therapy, intravenous infusions of high dose vitamin C as pro-oxidant, ozone, laetrile, multivitamin nutritionals, amino acids, and oral supplements. Emphasized is an acid-alkaline balance diet of organic food. Adjunct protocols include psychological counselling, breathing exercises, physical therapy/massage, detoxification strategies and nutritional education.

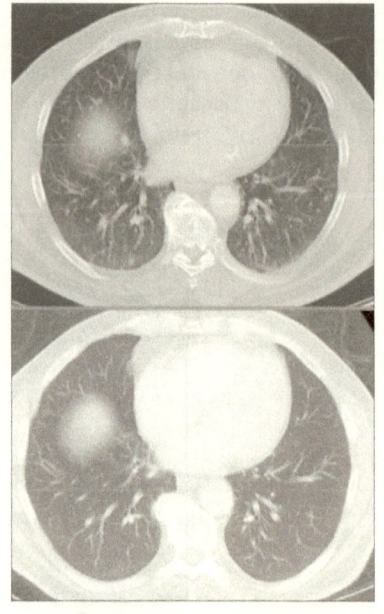

For a 76 year old man with a past medical history of diabetes mellitus II and hyperlipidaemia, in June 2008, diagnosis of melanoma involving the chest wall was confirmed by excisional biopsy with additional axillary node dissection revealing metastasis. His stated life expectancy without treatment was 1 year and 2 years with recommended treatments. Subsequent staging CT exam of chest revealed innumerable small nodules throughout both lungs consistent with metastatic melanoma. The patient was offered chemotherapy, however, due to concerns of possible side effects and low statistical chance for response, he elected to try an alternative immunotherapy program at the Issels Treatment Center. At the completion of the initial program, the patient underwent a monitoring CT scan which demonstrated complete resolution of all previously seen lung lesions, and unremarkable findings elsewhere.

Picture 17 (case 7): Top: Pre-therapy CT scan of chest with many nodules in bilateral lung fields consistent with metastatic melanoma.
Bottom: CT scan of chest performed 4 months since start of treatment with sustained remission of lung lesions.

After following the recommended home treatment protocol for 4 months, follow-up monitoring CT scan revealed continued unremarkable findings of the lung and elsewhere. In 2012 the patient's clinical status remained excellent as he continued the same regimen (see picture 17).

Case 8 and 9 - B-cell-lymphoma

The following four case descriptions are abridged and courtesy of Dr.M.Orange (Ita-Wegman Klinik, Arlesheim, Switzerland). At Ita-Wegman Klinik patients are treated for febrile induction therapy stationary for 1-2 weeks and as outpatients thereafter; this regimen is covered by private German and Swiss health insurances. These four treatments were done at the former Parc Attwood Clinic in Great Britain and published in 2010[101] and 2012[97].

Patient A

A 51-year-old female presented with two lesions on the left lower leg in May 2008. On presentation for mistletoe treatment in June 2008, the posterior lesion in the Achilles region measured 5 cm x 4 cm (picture 18) and the mid shin lesion measuring 4 cm x 2 cm, with a number of surrounding satellites, each < 1.5 cm. The 2 large lesions were raised, red, and warm to the touch. The overlying skin was thinned but intact. There were no signs of deep tissue infiltration; she reported no pain, and no neurological deficits were ascertained.

Histopathology had confirmed grade 1 follicular B-cell lymphoma. Staging computed tomography (CT) of chest, abdomen and pelvis reported no intra-abdominal or pelvic lymphadenopathy, but showed one 2.7 x 1.7 cm suspect inguinal lymph node. She had no history of injury or infection and no B symptoms, such as fatigue, night sweats, weight loss, or pruritus. She had longstanding lipoma's in different areas of the body but had been generally healthy.

Systemic immune-chemotherapy and involved-field radiotherapy was recommended, but the patient decided to keep this in reserve and first improve her immunity with mistletoe treatment.

Mistletoe treatment (MT, using Abnobaviscum® Fraxini), comprised a combination of intravenous (iv) intratumoural (it), and subcutaneous (sc) applications over 12.3 months and both iv and sc application for a further 8 months. Treatment was subdivided in an induction phase, wherein febrile reactions are elicited, and a post-induction (maintenance) phase.

MT was combined with whole-body hyperthermia - a technique to increase core temperature to 39° − 39.5°C for 2 - 4 hours with water-filtered infrared A radiation under controlled conditions. The skin lesions were not directly targeted. She had no other cancer treatments. Overall, the patient was treated at 64 different days with 133 mistletoe applications and 12 cycles of hyperthermia.

During induction, the patient had four febrile responses of >=38.5°C lasting <24 hours with maximum readings of 38.5°C (after treatment on day 3); 38.7°C (after day 4); 39.1°C (after day 8); and 38.6°C (after day 11). For the i.t. approach, the lesions were injected from the healthy skin margins to avoid breaking the paper-thin skin overlying the bulging lesions. The volume of injected mistletoe extract often exceeded 20 mL (20 mg/mL/ampoule) and was administered evenly intra- and perilesional while repositioning the needle during injection.

After i.t. application, the lesions swelled up immediate with prompt inflammation, followed by resolution. Over the course of time, the lesions became less inflamed.

The rate of regression seemed slightly accelerated after starting hyperthermia (day 37). After 4 months, there was a clear overall improvement. The lesions continued to show injection-associated fluctuations, and the posterior lesion was the first to resolve completely.

Fever was associated with grade 1 sickness and grade 2 - 3 fatigue. The maintenance i.v./i.t. administrations elicited fatigue (grade 1-2) for 1 - 3 days. No hypersensitivity was observed. S.c. injections elicited site responses of 4 – 5 cm erythema for less than two days. The i.t. injections with concentrated extract were uncomfortable with inflammatory response (erythema, swelling, tenderness) and transient increase of eodema of the lower leg lasting less than two days. Treatment comprised cooling applications, but did not require analgesia.

The patient described the treatment experience as follows:

> With the initial fevers and fluctuating energy levels, my treatment was intense, exhausting and it was the only thing I could do during that time: but not a burden and a meaningful experience. During one of the high fevers an old traumatic experience became disentangled and I have felt freed up since; I now feel better than before my cancer, physically and emotionally. I also felt empowered by the working together with my doctors to develop the best treatment for me. I am very grateful for my new health!

The lesions steadily decreased in volume, consistency, and redness. The remission was assessed by visual inspection and palpation and confirmed by three independent clinicians from three different clinical settings. The overall fitness and stamina of the patient improved. Re-scanning in May 2009 reported "No significant supraclavicular, axillary, mediastinal, retroperitoneal or pelvic lymphadenopathy; as before, there are inguinal nodes the largest of which is on the left and measuring 1.3 x 1.8 cm. This node was documented as 2.7 x 1.7 cm on staging." Haematology and biochemistry were normal. Given the favourable clinical signs of control, the i.t. injections were discontinued at 12.3 months. Combined i.v. and s.c. treatment and continued for another 8 months, and the lesions continued to regress. The areas blanched eventually, leaving depressed and hyper-pigmented areas (picture 18). Conventional therapy was deferred indefinitely. At last review in January 2013, the patient was doing well and remained in clinical remission.

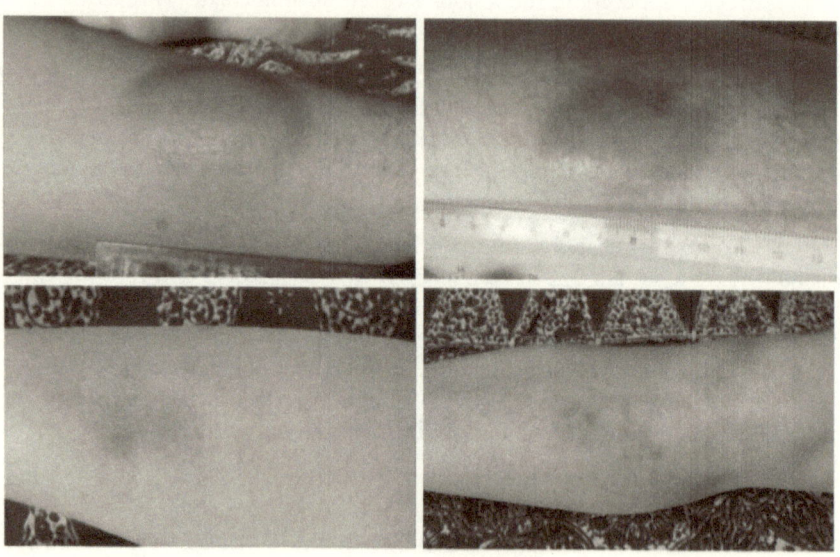

Picture 18 (case 8): Posterior lesion lower left leg (left) and anterior lesion lower left leg (right) between July 2008 (top) and May 2010 (bottom)

Patient B

A 52-year-old male was diagnosed with relapsing primary cutaneous marginal zone B-cell lymphoma (PCMZL). In December 2007, he developed a lesion in the left antecubital fossa (3 days after venipunction at the site), which was excised in May 2008. The histopathology showed nodal marginal zone lymphoma. A staging computer tomography scan of the neck, chest, abdomen, and pelvis showed no signs of systemic disease; trephine bone marrow biopsy, biochemistry, and haematology were normal.

Shortly after excision, the patient developed a second lesion at medial to the right anterior axillary fold of the chest wall. No other lesions were apparent, the patient was asymptomatic, had no recent weight loss, fatigue or night sweats. Several treatment options were recommended, including R-CVP, 10 fractions of involved-field radiotherapy of the 2 sites, or 6 months' pulsed chlorambucil; he declined these options.

The patient had a history of rosacea with keratitis; actinic keratoses of upper back (treated with cryotherapy); 2 basal cell carcinomas - one of the upper back and one of the left leg (both excised) and an uncertain diagnosis of facial cutaneous scleroderma with no visceral involvement, but raised titres of antinuclear factor, which was unresponsive to azathioprine and oral prednisolone (2004). He used nicotine and alcohol moderately; used no other regular conventional medications, had no formal allergies, and was mistletoe-naïve.

On presentation at August 2008, the patient was in good general health, with a Karnofsky Performance Scale status of >90%. The right upper thoracic lesion was 2 x 3 cm palpable and mobile. There were no other abnormal findings. The patient was counselled that an underlying autoimmune condition theoretically could be aggravated with MT. Combined i.v., i.t. and s.c. treatment (using Abnobaviscum® Fraxini) was given over 8.5 months, with 96 mistletoe applications all together. He received no other anticancer treatments.

He had 6 febrile responses (38° to 39.2°C) between days 5 and 87. After i.t. injections, the lymphoma lesion showed a similar response pattern of inflammatory swelling and erythema for up to 2 days, followed by slow resolution.

During the first month of MT the lesion increased to 4 x 5 cm, and then remained unchanged for 3 months. After the i.t. dose was increased to 100 mg, the lesion steadily diminished to become impalpable at 8.5 months. This complete response (CR) was clinically verified by three clinicians in two separate institutions. The i.t. treatments were discontinued in April 2009, and i.v. and s.c. treatments were continued until November 2010. A CT scan in March 2010 was unremarkable.

During the first 3 months, treatment was challenging. The fever episodes in particular were accompanied by sickness and grade 1 - 2 fatigue. Once, grade 2 phlebitis developed at an i.v. cannulation site and resolved spontaneously; interestingly, no local relapse resulted from this. The s.c. and i.t. doses were followed by typical inflammatory site reactions that resolved without scarring or subcutaneous fibrosis. No hypersensitivity and no signs or symptoms of autoimmune reactivation were observed. After six months, the patient consistently reported improved vitality and well-being. The patient was last reviewed in December 2012 and was doing well and in remission; no new lesions had developed.

Case 10 - Merkel cell carcinoma

Merkel cell carcinoma (MCC) is a rare and aggressive skin cancer with a 30% mortality that occurs in persons over the age of 50 or who are immunosuppressed. Most lesions occur in the head and neck region and on sun-exposed extremities and are often found near to or co-existent with other lesions of sun damaged skin, including Bowen disease. Primary excision with wide margins is the treatment of choice, together with elective lymphadenectomy or sentinel node biopsy. MCC has a high recurrence rate and adjuvant radiotherapy may improve local control, but without survival advantage, and routine use of chemotherapy is not recommended.

A 75-year-old female had a 6-8 cm MCC excised from the right upper arm in September 2007. The surgical margins were clear and CT-scan staging showed no evidence of macroscopic spread. In December 2007 the patient relapsed with a right axillary lymph node measuring 3 cm on ultrasound (US). A CT-scan confirmed the lesion, showed no distant spread and routine blood work was normal. She had no allergies and took no regular medications. In addition, she was diagnosed with a rectal carcinoma in December 2008, following a 12-month history of intermittent rectal bleeding. On presentation for treatment in February 2008, the patient was in good condition, Karnofsky Performance Score >90%. The MCC was treated with mistletoe therapy, in a combination of i.v., i.t. and s.c. administration. With occasional exceptions, only Abnobaviscum® Fraxini was used. At induction, combined doses were escalated from 42.2 mg to 100 mg and 162 mg within the first 3 days, in favour of i.v. and i.t. administration – which were followed by marked local and febrile reactions. Initially, the axillary tumour enlarged to measure 4 cm × 5 cm and thereafter a steady reduction was noted with the lesion becoming impalpable after four months. Subsequently, combined i.v./i.t. administrations were continued at monthly intervals, whilst s.c. injections continued with 4 mg twice a week. Altogether, she received 139 mistletoe applications over 9.8 months. The regression of the node was independently confirmed by two teams.

In January 2009 the patient had surgery for a rectal adenocarcinoma, which was incompletely excised with involved surgical margins. She has made a good recovery and declined adjuvant chemotherapy. She remains in clinical remission of MCC, confirmed by CT-scan in March 2009. Regular s.c. and i.v. mistletoe treatment continued. In January 2013 the patient was well and asymptomatic of both MCC and rectal carcinoma. By RECIST criteria this outcome is consistent with a complete clinical response.

CANCER FEVER THERAPY: CASES

Case 11 - Bilateral breast cancer

In September 2004 a female patient was diagnosed with a 3.8 mm breast cancer, confirmed by biopsy as grade 1 invasive ductal, oestrogen sensitive carcinoma. She declined staging and primary surgery. In August 2005 this lesion had increased to 5.4 mm. In December 2005 the right breast lesion was 5 mm and in addition, she had a new 23 mm lesion of the left breast, confirmed malignant by core-biopsy: grade 3 invasive ductal, both oestrogen and progesterone receptor negative, and HER2-positive. There was no regional lymphadenopathy and no clinical signs of distant disease. She declined further staging investigations and declined treatments. Her hospital team had recommended neo-adjuvant chemotherapy, bilateral mastectomy, further adjuvant chemotherapy, radiotherapy - followed by trastuzumab and 5 years hormone treatment. However, the patient felt that such treatments would be deleterious to her health and immune system. After thorough discussions at Park Attwood Clinic, it was agreed to continue follow-up at her hospital and to keep primary surgery as an option. Informed consent was based on the understanding that mistletoe treatment would not replace recommended therapies. On presentation for VAE treatment in January 2006 she was asymptomatic, Karnofsky Performance Score 100%. She had silicone implants at age 28, no current co-morbidity, reported no allergies, took no regular conventional medications and was mistletoe naïve. She had started taking an ayurvedic herbal remedy (Carctol, from Cankut Ltd.).

The patient had a total of 31 months of mistletoe treatment with neo-adjuvant intent. She developed marked (grade 2) inflammatory site reactions of skin (abdomen) and left breast following the i.t./s.c. administrations on days 1, 5 and 16 were marked; these settled over 3–5 days with cooling local applications alone and required no analgesia or antipyretics. The systemic responses comprised mild (grade 1) chills and rigors, moderate (grade 2) fever (<40°C) and severe (grade 3) fatigue for 2 days, with moderate fatigue (grade 2)-requiring treatment free recovery of 3, 11 and 6 days, respectively. After 6 weeks her resilience and stamina had improved and the responses to subsequent i.t./s.c. administrations lessened to mild (grade 1) fatigue for no more than 2 days. In spite of substantial increase of doses for i.t. injections, the site responses were mild (grade 1) for 3–5 days. S.c. injections elicited mild site reactions of 4–5 cm erythema, lasting less than 2 days. No hypersensitivities were observed.

The left breast tumour received a total of 40 i.t. applications, alongside 129 weekly s.c. injections. Post-induction, the intervals between the i.t. injections were 1 and 2 weeks during the first 4 months and gradually extended to 2, 4 and 6 weeks, according to tumour response and patient tolerance. Although the right breast tumour was not injected, it was the first to regress and became impalpable during the first month of treatment and US-negative after 5.5 months. The left breast tumour was impalpable at 14 months and US-negative at 31.2 months, representing a complete clinical and radiological response. The patient discontinued mistletoe treatment after 2.6 years and declined surgical closure because she felt that her breast cancer was cured. In January 2010 the patient was well and in durable clinical remission.

CANCER FEVER THERAPY: CASES

Case 12 - Breast and mesothelioma cancer

Klinik in Arlesheim near Basel, Switzerland, is one of the clinics with extensive experience in mistletoe therapy. The initial manufacturing steps of Iscador, a mistletoe drug branded by Weleda, are prepared at the Verein für Krebsforschung, located at the Lukas-Klinik. Iscador, in general, is applied subcutaneously, which does not induce fever at normal dosage or, with high dosage, only during the first four weeks of application. I.v. applied Iscador usually induces a temperature elevation of > 39°C for about 2 hours and >38°C for another two hours. If Iscador is applied i.v. in the morning, body temperature usually is down to normal in the evening.

A female patient, born 1952, was diagnosed with right breast carcinoma. The tumour was dissected together with 20 positive lymph nodes in July 1992, followed by radiation of the right breast and shoulder region. Adjuvant chemotherapy or hormone therapy were recommended but rebuffed by the patient.

In December 1992 the patient started a monthly i.v. Iscador therapy. In December 1998 an invasive peritoneal methotheliom was diagnosed, a rare form cancer often caused by asbestos inhalation. She presented with ascites, an accumulation of fluid in the peritoneal cavity. Chemotherapy was recommended and declined again, rather Iscador treatment was increased to a higher dosage and hyperthermia treatment started in January 1999. In February 1999 ascites could not be confirmed any more. The tumour was removed by surgery and histologically confirmed in October 1999. In July 2001 no nodes were found by MRT. Slightly enlarged ovaries were revealed by CT in October 2001 and surgically removed the same month. New tumour masses were detected in October 2002. The patient intensified infusions, inducing fevers up to 40°C, whereupon nodes regressed.

The patient consequently continued Iscador infusions and hyperthermia, in recent years in 2-3 month intervals. In April 2013 she reported best quality of life.

Case 13 - Colon carcinoma with hepatic metastasis

The private Hufeland-Klinik in Bad Mergentheim, Germany was founded by one of Issels' pupils, Dr. Wolfgang Wöppel, and combines several measures into a "holistic immunebiological therapy". For each patient a therapy portfolio is arranged individually from colon hydro therapy, colon sanitation, ozone therapy, self-blood therapy, thymic peptides, high dosage vitamines, hyperthermia and fever therapy.

Patient B., born 1943 was diagnosed with colon carcinoma, which had already metastasized into the liver. Between July 1994 and July 1995 he received palliative chemotherapy, which led to partial regression of liver metastases. In April 1995 he started adjuvant therapy at Hufeland. In September 1995 not a single primary or secondary neoplasm could be detected by ultrasound and CT. The primary clinic in Erlangen was so astonished that they intended to "cut him open for looking", because they could not believe CT and ultrasound findings. The patient commenced the Hufeland therapy until 2006 and once in a while in subsequent years. Complete remission was confirmed in July 2012.

Case 14 - Metastasing melanoma

For the following case no detailed patient records were available. It is included here, because it was broadcasted with a patient interview in a TV-show about "Wunderheilungen" (magic healings) in 2010. Fever therapy was not mentioned in this show.

Patient S., born 1955, was diagnosed with malignant melanoma in 1991. In 1992 new skin lesions were detected as well as lymph node metastases. He was terminally ill after two surgeries in 1992. Physicians offered a very aggressive chemotherapy in a clinical trial but essentially could give him no hope. In April 1992 he went into Hufeland-therapy for eight weeks, including fever therapy using bacterial extracts. During this time, metastases continued to grow. One superficial thigh lesion had the size of a large hand. Afterwards the patient continued fever therapy in an ambulant setting. Remission set in early 1993. The family physician noted, almost hesitantly, "I cannot but admit that node sizes decrease." Complete remission was diagnosed after three month.

Case 15 - Melanoma relapse

„Klinik im LEBEN" in Greiz, Thüringen, is one of the German private clinics offering active fevertherapy, mistletoe therapy, hyperthermia and their combination. „Klinik-im-LEBEN" tries to leave behind the application of bacterial extracts and rather use applied drugs containing bacterial or viral substances designating fever as a frequent adverse effect in the instruction leaflet (see table 3). First results are encouraging with respect to fever induction.

Patient W.M. was treated on melanoma by surgery and interferone. 2004, aged 57, a relapse was found under the left armpit (stadium T3 N1 Mx). The patient refused another surgery. The first fever therapy in May 2004 shrank the relapse after three fever inductions. After additional eight fever strikes the tumor shrank down to a small residual . Last fever strike was in May 2007. As a surprising side effect Morbus Bechterew pain persistent since years vanished.

CANCER FEVER THERAPY: CASES

Case 16 - Mamma carcinoma with bone metastases

Patientin K.A., born 1957, presented herself in March 2014 with invasive ductal lobular mamma carcinoma (stadium L1 V0 R1 M1 G2) and bone metastases distributed over the skeleton, leading to severe permanent pain. After three weeks of bi-weekly fever and hyperthermia treatments, pain was considerably reduced. Treatment was continued in monthly intervals. CT was constant in February 2015, no pain was reported. The patient felt much better, debitable and able to work again. From summer 2015 treatment continues in bi-monthly intervals.

Cases - wrap up

There is no unequivocal treatment regimen which can be abstracted from theses cases. All of these treatments were not standardized and based either on long experiences, supportive evidence from historical treatments or more or less plausible assumptions. There are neither pre-clinical studies nor human trials available which would obey evidence based standards. But from these and other cases one is tempted to conclude that PRRL-therapy can heal. Factoring in our immunological considerations it might help if

- the bulk of tumour load is removed by surgery
- the immune system is not too badly damaged by prior immune compromising therapy
- both the patient and the treating physician have a strong confidence into the power of the immune system
- drugs containing PRRL-substances are applied metronomically, at least in the initial phase as frequent as possible
- a long treatment running over weeks and month is anticipated
- both patient and physician are aware, that while general well-being is often reported to improve soon, more distinct signs of therapy success like softening or shrinking of lesions often can be seen soon but also may occur with a time lag of some weeks
- fever is appreciated rather than suppressed
- herds of chronic inflammation, e.g. dead teeth, are removed
- supportive measures like regular exercise, healthy nutrition and removal of sources of stress are implemented

The lack of official clinical trials can make financing by health insurances difficult. For instance, health insurances in Germany compensate treatments in a private clinic, but only for a limited amount of time. It should be underlined that once the body got used to frequent body temperature elevations, prolongation of treatment in an ambulant setting - which is much cheaper - should be no problem, from a medical point of view.

Both patient and doctor should know that, although general well being often impoves upon fever therapy, softening or size reduction of a solid lesion often occur only with some delay, which can last few weeks. Even a transient size increase can be seen, perhaps caused by immune cell influx into the tumour bed. Physicians with fever therapy experience judge treatment success in the beginning by general condition rather than tumour size.

Even under optimal fever therapy exaggerated hopes are misplaced. Coley's many case studies have shown that tolerance against the neoplastic growth cannot be broken in some patients, for unknown reasons.

Fever therapy is not approved by health insurances. In Germany, stationary mistletoe treatment - which arguably should be augmented by other PRRL drugs - is refunded for one to two weeks only. Since a longer metronomic treatment is desirable, a subsequent ambulant treatment should be considered, where the patient comes into the clinic in the morning and leaves in the evening. Refreshment treatments in longer intervalls might be desirable.

Prophylaxis and aftercare

Any cancer that can be avoided statistically counts as a healing. Lowering cancer incidence translates directly into lowering cancer mortality. PRRL substances, in perspective, might also play an important role in cancer prophylaxis [102].

The main sub-population to be targeted by prophylactic PRRL-treatment are treated cancer patients who look for means to reduce the risk of relapse. Another clientele might be people who for a long time did not experience a feverish infection. These might be treated by a series of prophylactic PRRL vaccinations. But beware: from the beginning of the book up to this point we carefully tried to stick to scientific facts. From here on we slightly move into the realm of speculation.

In the public and also among physicians we find a persistent myth that cancer is a disease that breaks out and than inevitably develops rapidly. Severe stress and traumata are suspected to be inducers of such an outbreak. But does every traumatised patient get cancer ? Under this paradigm, how can we explain certain mouse strains which predictably develop cancer within one year despite a wonderful stressless life ? How can one explain the mantra of oncologists "my patients were never sick"?

The majority of cancers develop slowly, starting with cells carrying only few genetic aberrations up to the hundreds of genetic defects we find in clinically evident tumours. Cancer develops as a stochastic process usually running over many years.

Smoking as a common societal habitude developed in the US after world war II soldiers had been equipped with daily packets of cigarettes. The number of cigarettes consumed in the US by men increased with a steep slope in the years after the war. Lung cancer incidence rose with an exactly identical slope - more than 30 years later[2].

> One of the physicians who initially rebuffed the model of a silently lingering disease, later wrote me a note: "Looking backwards and after some reflection I must confess that many of my cancer patients had no or only few feverish infections during childhood."

PROPHYLAXIS AND AFTERCARE

The immune system most likely is capable to keep most aberrant cells in check for a long time, and the control of these elopers by surrounding tissue is not completely lost. Cancer cells in early stages obviously can be removed by feverish infections. The epidemiologic data in table 1 cannot be explained otherwise. The observation that cancers often become dangerous after severe stress or trauma is not incompatible with this hypothesis. During stress and trauma the strength of the immune system declines, the immunological shield becomes weaker, and once the tumour has reached a certain size it cannot be kept in check further.

How strong could the prophylactic protection of a feverish infection be?

Kölmel et al. compared 603 melanoma patients with a similar control group. All participants were interviewed about number and severity of feverish infections in the past. It turned out, the more feverish infections were reported, the lower was the risk to develop melanoma later. With three and more hefty feverish infections during the past years the risk to develop melanoma was reduced by forty percent[72]. Post-operative infections provide a similar risk reduction (see table 1).

This amount of risk reduction is comparably enormous, because we have to compare with the reduction in cancer mortality we have achieved so far by standard therapy.

Cancer mortality is one benchmark to judge a cancer therapy. Another, more frequently applied measure is 5-year survival. This statistic tells how many people live five years after cancer diagnosis. 5-year survival rates are biased by our technical improvements in diagnosis, leading to earlier diagnosis and higher 5-year survival rates compared to fifty years ago. Some part of increased survival time probably is indeed attributable to the reduction of tumour load induced by current drug therapies and surgery. Of course a longer life is a win for the individual patient. But it does not tell much about our healing success on average, because she or he still may die from cancer later.

The most honest measure of success in our battle against cancer is cancer mortality. How much were we able to reduce the fraction of people in the population dying from cancer?

Despite all our immense financial and intellectual efforts in science and medicine we were able to reduce cancer mortality by only 5%-10% during the last 50 years. And this reduction is still biased by reduced smoking. In other words, if smoking habitudes had been constant over the last 50 years, we could not even claim those meager 5%-10% reduction in cancer mortality.

If we could propagate the above mentioned 40% reduction in cancer incidence to other forms of cancer as well, the prophylactic potential of infections - or PRRL-therapy for that matter - is nothing else but tremendous.

Therefore we should carefully reflect whether it is wise to wipe any childhood disease by vaccination, to shoot down any little infection by antibiotics, to suppress any fever immediately by drugs or suppository.

Suggestions like "live less sterile and hygienic" or "permit yourself a couple of uneasy days in bed without antibiotics" may raise the blood pressure of your physician and still be serious. Because it is imperative to valuate all data without pre-occupation and take a fresh look to decide which is better: the quick relief from an unpleasant infection now or a certain protection from cancer later. In our jobs we are asked to be functional soon. But we also want to be functional for long.

PRRL-vaccinations will differ from normal vaccinations. To resemble a proliferative infection it will be needful to apply a series of injections over the course of some days. This form of vaccination could demand about the same time as a short fasting period. It might be most interesting for cancer patients who have left behind the trauma of diagnosis and first therapy, who had some time to regenerate their immune system, and who now look for a way to reduce the likelihood of relapse and, about once a year, are willing to invest one or two weeks.

Diet and cancer

To produce energy, our minds and our cancer cells prefer one particular class of molecules over all others: carbohydrates, in particular sugars. Besides pure saccharose and glucose in candies and cakes the most important suppliers of carbohydrates in our nutrition are bread, potatoes, noodles, rice, which can easily be digested into small fuel molecules like glucose. Since one of these foodstuffs is part of almost every meal, our blood sugar levels are always up.

This is a new situation for our species. Over million years, with the exception of the last 10000 years, our ancestors were hunter-gatherers. They used to stray out into the woods to collect leafs, fruits and nuts and hunt some wild game. Later, with the invention of the spear about 300000 years ago, nutrition contained larger portions of protein and fat. There is not much carbohydrate in such a diet. Presumably, there were days with little or without food. Farming, planting of cereals, domesticating aurochsen to cattle, all that was invented only ten thousand years ago. Over the vast part of our evolutionary history, not a blood sugar level permanently in the sky was normal, but rather long periods of low blood sugar levels, which sometimes might rise a bit when a bunch of sweet fruits was digested.

We know that some limited evolutionary adaptation can occur within such a tiny period of ten thousand years. The ability to digest milk as an adult, for instance, happened during this time and spread into the genomes of many - but not all - humans since. Roughly 25% of all people can digest milk sugar (lactose), and about half of all people can digest it partially. Before, milk was food for human babies exclusively and remains baby food for all mammals including 25% of our own species. Genetic adaptation is slow.

Did we adapt enough to cope with the drastic shift towards a carbohydrate rich diet since ? Amylase in saliva is an enzyme to digest starch into smaller sugars. Few individuals have more than one gene for amylase and have higher amounts of the enzyme in saliva. They digest starch more rapid and more efficient and raise their blood insulin levels faster [103]. Do these people get less diabetes ? Is this an adaptation to fully cope with our sudden carbohydrate-rich diet ? We do not know. What we know is that both the consumption of sugars and the incidence of diabetes skyrocket, compared to all previous generations in human history.

Meanwhile we have preliminary hints that low or at least alternating blood sugar levels not only are not disadvantageous, but have serious positive effects on metabolism and health, including epilepsy, diabetes, Alzheimers, Parkinson, cancer[104,105].

10-12 hours after the last meal blood sugar is used up. The body starts to produce glucose from glycogen depots in liver and muscles. With on-going carbohydrate depletion the body turns towards burning body fat, which is in full swing after 3-4 days. An intermediate product of fat degradation are so-called ketone-bodies, which the brain and muscles are happy to burn instead of glucose. The ravenous appetite for sweets declines. Appetite becomes more stable, the many walks to the freezer over the day take off.

Low-carb diet exists in more or less restrictive flavours. In the stricter form even fruits with higher carbohydrate content like fig, date, banana are avoided., while in relaxed versions of low-carb diet those are allowed. In any case, white industrial sugar with direct access to the blood is reduced to the minim.

There is no risk with low-carb. Normal weight people without problem can fast for several days and permanently live on low carb. A minimal amount of carbs needed by the body for red blood cells is available in vegetables or, during fasting, can be produced by the liver from body fat. We are built for living well without or little carb.

Upon switching to burning fat, several hormonal changes can be observed. One of them, brain derived neurothrophic factor, raises up to fourfold. This hormone is capable to stimulate growth of new nerve cells and protects from Azheimers and Parkinson disease. Indeed, fasting mice are more active than their satiated peers. Another hormone, IGF-1, goes down to reach its deepest level after 3-4 days of fasting. The same for insulin, which is related to IGF-1. High levels of IGF-1 and insulin are known to be correlated with a higher risk for diabetes and cancer. Lower levels of both IGF-1 and insulin during a carbohydrate-low diet decrease the risk to develop diabetes and cancer.

The effect of fasting with respect to cancer is both prophylactic and therapeutic, just like an acute infection. Two days of fasting slowed tumour growth in mice, with additional fasting periods pronouncing the effect. The authors of this study presume similar outcomes in humans[106].

Fasting appears to be a bigger threat to cancer cells than to normal cells. Remember: cancer cells always carry many genetic defects, at least dozens and usually hundreds. Cancer cells have optimized themselves to divide fast and independent from surrounding tissue, but are less capable to perform many other metabolic tasks compared to normal cells. The advantage of rapid cell division can turn into a disadvantage, if adaptation to new "environmental" conditions is needed. Fasting - which in this respect might be resembled by a poor-carb diet - is a new environmental condition, since to cope with changed energy supply, activation of a separate repertoire of genes is required. The change of multiple environmental conditions at the same time should be an even greater threat to cancer cells. This has already been shown for the combination of fasting with chemotherapy or radiotherapy: this double-hit led to longer survival of more cancer mice even with glioma, one of the most aggressive brain tumours[107]. Breast cancer patients with high insulin levels have a double risk of relapse, compared to patients with low insulin levels[108]; higher blood sugar levels raise the risk to develop cancer[109].

We can only speculate how cancer cells might suffer from the double-hit fever plus low-carb.

Vista

The following observations point into the same direction: a correlation between feverish infection and spontaneous remission; lower cancer risk in people with a personal history of feverish infections; the old clinical experiences made by Coley and Klyuyeva; mistletoe therapy. Fever therapy presumably has unleveraged potential in the treatment of cancer, *when* aplied correctly. Correctly means under synergistic immune stimulatory agency of a PRRL *cocktail*, applied *long* enough in *short* intervalls. If several fever inductions per week are not possible, for instance, because the patient is too weak, intermittend application of a subfebrile and less exhausting dosage will be better than pausing treatment, to keep the innate immune system from falling back into dormancy.

Prejudice against induction of fever may be broadly and deeply ingrained, but still be unjustified. Fever induced by PRRL is not a fever caused by proliferating pathogens, but will invariably decline in short time. Compared to standard cancer therapy, fever therapy does not cause long lasting or by itself cancer inducing harm. Treatment costs are comparably low.

Although chemotherapy and radiation, due to their immune compromising effects, should nor run parallel with fever therapy, both options are not necessarily incompatible when applied in succession. For many forms of cancer there should be enough time to start a fever therapy before surgery and continue for a while after surgery, before more drastic measures are considered. Issels' observation, hyperthermia reduces chemo therapy side effects or can be used for dosage reduction, might be taken into consideration. If chemotherapy or radiotherapy are felt both by physician and patient to be required for debulking, dosage should be kept at a minimum and fever therapy could be taken up again after the immune system has recovered to some extent.

Mistetoe therapy, which at the beginning can cause fever, is based on a tremendous foundation of experience in Europe, generated over decades. Mistletoe extract probably is a weaker innate immune stimulant compared to bacterial extracts. But there are several approved drugs on the market, which, according to their respective instruction leaflet, contain PRRL and where fever is reported as common side reaction (table 3). These could be tested to augment mistletoe extract: mistletoe therapy 2.0. Without approvement hassle[i], at low expense for the combination. Not only as adjuvant or palliative but primary treatment. Here and today.

[i] Wikipedia: "[German] Medical Leitlinien [guidelines which recommend chemotherapy for many forms of cancer] are ... statements to support physicians... and patients with their decisions. They are - other than Richtlinien [directives] - not binding and have to be adjusted to the individual case."

References

1 Hall, S. A commotion in the blood. (Owl Publishing, 1998).
2 Weinberg, R. A. The Biology of Cancer. (Garland Science, 2007).
3 Busch, W. Aus der Sitzung der medicinischen Section vom 13.November 1867. Berliner Klinische Wochenschrift 5, 137 (1868).
4 Fehleisen, F. Über die Züchtung der Erysipelkokken auf künstlichem Nährboden und ihre Übertragbarkeit auf den Menschen. Deut.Med.Wochenschr. 85, 553-554 (1882).
5 Roger, G. H. Seances et Mem Soc de Biol Paris 2, 573-580 (1890).
6 Coley, W. B. The Treatment of Malignant Tumors by Repeated Inoculations. Am. J. Med. Sci. 105, 487-511 (1893).
7 Coley-Nauts, H. C., Fowler, G. A. A. & Bogatko, F. H. A review of the influence of bacterial infection and of bacterial products (Coley's toxins) on malignant tumors in man. Acta Med.Scand. 145, 5-102 (1953).
8 Christian, S. & Palmer, L. An apparent recovery from multiple sarcoma with involvement of both bone and soft parts treated by toxin of erysipelas and bacillus prodigiosus. Amer.J.Surg. 43, 188-197 (1928).
9 Nauts, H. C. & McLaren, J. R. Coley toxins--the first century. Advances in experimental medicine and biology 267, 483-500 (1990).
10 Wiemann, B. & Starnes, C. O. Coley's toxins, tumor necrosis factor and cancer research: a historical perspective. Pharmacol Ther 64, 529-564 (1994).
11 Klyuyeva, N. G. & Roskin, G. I. Biotherapy of malignant tumours. (Pergamon Press, 1963); http://bioinfo.tg.fh-giessen.de/cancer/klyuyeva-1963.pdf
12 Everson, T. C. & Cole, W. H. Spontaneous regression of cancer. (J.B.Saunders & Co, Philadelphia, 1968).
13 Rohdenburg, G. Fluctuations in the growth energy of malignant tumors in man, with especial reference to spontaneous recession. J Canc Res 3, 193-225 (1918).
14 Hobohm, U. Fever and cancer in perspective. Cancer Immunol Immunother 50, 391-396 (2001).
15 Diamond, L. K. & Luhby, L. A. Pattern of 'spontaneous' remissions in leukemia of the childhood, observed in 26 of 300 cases. Am.J.Med. 10, 238ff (1951).
16 Stephenson, H. E., Jr. et al. Host immunity and spontaneous regression of cancer evaluated by computerized data reduction study. Surg Gynecol Obstet 133, 649-655 (1971).

REFERENCES

17 Black, M. S., Opler, S. & Speer, S. Structural representations of tumor-host relationships in gastric carcinoma. Surg.Gynec.Obstet. 102, 599-603 (1956).

18 Alexander, P., Eccles, S. & Gauci, C. The significance of macrophages in human and experimental tumours. Ann NY Acad Sci 276, 124-133 (1976).

19 Pierce, E. The 34th Rovenstine Lecture ; http://www.apsf.org/about/rovenstine/part4.mspx

20 Vestergaard, H., Westergaard, T., Wohlfahrt, J., Hjalgrim, H. & Melbye, M. Tonsillitis, tonsillectomy and Hodgkin's lymphoma. Int J Cancer 127, 633-637, doi:10.1002/ijc.24973 (2010).

21 Lown, B. The lost art of healing. (Ballantine Books, 1999).

22 Ridker, P. M. et al. Rosuvastatin to prevent vascular events in men and women with elevated C-reactive protein. N Engl J Med 359, 2195-2207, doi:10.1056/NEJMoa0807646 (2008).

23 Wolmark N, W. H., Hyams DM. Randomized Trial of postoperative adjuvant chemotherapy with or without radiotherapy for carcinoma of the rectum. J Natl. Canc. Inst. 92, 388-396 (2000).

24 Nagakawa T, N. M., Futakami F. Results of extensive surgery for pancreatic carcinoma. Cancer 77, 640-645 (1996).

25 Gudjonsson, B. Critical look at resection for pancreatic cancer. Lancet 348 (1996).

26 Gotzsche, P. C. & Nielsen, M. Screening for breast cancer with mammography. Cochrane Database Syst Rev, CD001877, doi:10.1002/14651858.CD001877.pub2 (2006).

27 Humphrey, L. L., Helfand, M., Chan, B. K. & Woolf, S. H. Breast cancer screening: a summary of the evidence for the U.S. Preventive Services Task Force. Annals of internal medicine 137, 347-360 (2002).

28 Woloshin, S. & Schwartz, L. M. The benefits and harms of mammography screening: understanding the trade-offs. JAMA 303, 164-165, doi:10.1001/jama.2009.2007 (2010).

29 Kalager, M., Adami, H. O., Bretthauer, M. & Tamimi, R. M. Overdiagnosis of invasive breast cancer due to mammography screening: results from the norwegian screening program. Annals of internal medicine 156, 491-499, doi:10.1059/0003-4819-156-7-201204030-00005 (2012).

30 Zahl, P. H., Maehlen, J. & Welch, H. G. The natural history of invasive breast cancers detected by screening mammography. Archives of internal medicine 168, 2311-2316, doi:10.1001/archinte.168.21.2311 (2008).

31 Jorgensen, K. J., Klahn, A. & Gotzsche, P. C. Are benefits and harms in

mammography screening given equal attention in scientific articles? A cross-sectional study. BMC medicine 5, 12, doi:10.1186/1741-7015-5-12 (2007).

32 Giuliano, A. E. et al. Axillary dissection vs no axillary dissection in women with invasive breast cancer and sentinel node metastasis: a randomized clinical trial. JAMA 305, 569-575, doi:10.1001/jama.2011.90 (2011).

33 Andriole, G. L. et al. Mortality results from a randomized prostate-cancer screening trial. N Engl J Med 360, 1310-1319, doi:10.1056/NEJMoa0810696 (2009).

34 Welch, H. G. & Albertsen, P. C. Prostate cancer diagnosis and treatment after the introduction of prostate-specific antigen screening: 1986-2005. J Natl Cancer Inst 101, 1325-1329, doi:10.1093/jnci/djp278 (2009).

35 Topol, E. The creative destruction of Medicine. (Basic Books, 2012).

36 Peterson, R. D. A., Kelley, W. D. & Good, R. A. Ataxia-telangiectasia. Its association with a defective thymus, immunologic deficiency disease, and malignancy. Lancet 1, 1189-1193 (1964).

37 Kuppen PJK. The infiltration of experimentally induced lung metastases of colon carcinoma CC531 by adoptively transferred Interleukin-2 activated natural killer cells in Wag rats. J.Canc. 56, 574-579 (1994).

38 VanderBruggen, P. et al. A gene encoding an antigen recognized by cytolytic T- lymphocytes on a human melanoma. Science 254, 1643-1647 (1991).

39 Rosenberg, S. A., Yang, J. C. & Restifo, N. P. Cancer immunotherapy: moving beyond current vaccines. Nature medicine 10, 909-915 (2004).

40 Dunn, G. P., Old, L. J. & Schreiber, R. D. The immunobiology of cancer immunosurveillance and immunoediting. Immunity 21, 137--148 (2004).

41 Yang, Y., Huang, C. T., Huang, X. & Pardoll, D. M. Persistent Toll-like receptor signals are required for reversal of regulatory T cell-mediated CD8 tolerance. Nat Immunol 5, 508-515 (2004).

42 Trieb, K., Sztankay, A., Amberger, A., Lechner, H. & Grubeck-Loebenstein, B. Hyperthermia inhibits proliferation and stimulates the expression of differentiation markers in cultured thyroid carcinoma cells. Cancer Lett 87, 65-71 (1994).

43 Maletzki, C., Linnebacher, M., Savai, R. & Hobohm, U. Mistletoe lectin has a shiga toxin-like structure and should be combined with other Toll-like receptor ligands in cancer therapy. Cancer Immunol Immunother, doi:10.1007/s00262-013-1455-1 (2013).

44 Beuth, J. et al. [Immunoactive action of mistletoe lectin-1 in relation to dose]. Arzneimittel-Forschung 44, 1255-1258 (1994).

45 Park, H. J. et al. TLR4-mediated activation of mouse macrophages by

REFERENCES

Korean mistletoe lectin-C (KML-C). Biochemical and biophysical research communications 396, 721-725, doi:10.1016/j.bbrc.2010.04.169 (2010).

46 Kienle, G. S., Glockmann, A., Schink, M. & Kiene, H. Viscum album L. extracts in breast and gynaecological cancers: a systematic review of clinical and preclinical research. Journal of experimental & clinical cancer research : CR 28, 79, doi:10.1186/1756-9966-28-79 (2009).

47 Solomayer, E. F. et al. Influence of adjuvant hormone therapy and chemotherapy on the immune system analysed in the bone marrow of patients with breast cancer. Clin Cancer Res 9, 174-180 (2003).

48 Mackall, C. L. T-cell immunodeficiency following cytotoxic antineoplastic therapy: a review. Stem Cells 18, 10-18, doi:10.1634/stemcells.18-1-10 (2000).

49 Deidier, A. Dissertation Medecinal et Chirurgical sur les Tumeurs, Paris, (1725).

50 Pearl, R. Cancer and tuberculosis. Am J Hyg 9, 97-162 (1929).

51 Braunstein, A. Krebs und Malaria. Zeitschrift fuer Krebsforschung 29, 330-333 (1929).

52 Braunstein, A. Experimentelle und klinische Grundlagen fuer Malariabehandlung des Krebses. Zeitschrift fuer Krebsforschung 29, 468-490 (1929).

53 Engel, P. Ueber den Infektionsindex der Krebskranken. Wiener Klinische Wochenschrift 47, 1118-1119 (1934).

54 Sinek, F. Versuch einer statistischen Erfassung endogener Faktoren bei Carcinomerkrankungen. Z Krebsforsch 44, 492-527 (1936).

55 Kienle, G. S. Fever in cancer treatment: Coley's therapy and epidemiological observations. Global advances in Health and Medicine 1, 90-98 (2012).

56 Witzel, L. [History and other diseases in patients with malignant neoplasms]. Medizinische Klinik 65, 876-879 (1970).

57 Zygiert, Z. Hodgkin's disease: remissions after measles. Lancet 1, 593 (1971).

58 Ruckdeschel, J. C., Codish, S. D., Stranahan, A. & McKneally, M. F. Postoperative empyema improves survival in lung cancer. Documentation and analysis of a natural experiment. N Engl J Med 287, 1013-1017, doi:10.1056/NEJM197211162872004 (1972).

59 Newhouse, M. L., Pearson, R. M., Fullerton, J. M., Boesen, E. A. & Shannon, H. S. A case control study of carcinoma of the ovary. British journal of preventive & social medicine 31, 148-153 (1977).

60 Remy, W. et al. Tumorträger haben selten Infekte in der Anamnese. Med

Klinik 78, 95-98 (1983).
61 Ronne, T. Measles virus infection without rash in childhood is related to disease in adult life. Lancet 1, 1-5 (1985).
62 Enterline, P. E., Sykora, J. L., Keleti, G. & Lange, J. H. Endotoxins, cotton dust, and cancer. Lancet 2, 934-935 (1985).
63 van Steensel-Moll, H. A., Valkenburg, H. A. & van Zanen, G. E. Childhood leukemia and infectious diseases in the first year of life: a register-based case-control study. American journal of epidemiology 124, 590-594 (1986).
64 Grossarth-Maticek, R., Frentzel-Beyme, R., Kanazir, D., Jankovic, M. & Vetter, H. Reported herpes-virus-infection, fever and cancer incidence in a prospective study. Journal of chronic diseases 40, 967-976 (1987).
65 Rotoli, B., Formisano, S., Martinelli, V. & Nigro, M. Long-term survival in acute myelogenous leukemia complicated by chronic active hepatitis. N Engl J Med 307, 1712-1713, doi:10.1056/NEJM198212303072721 (1982).
66 Treon, S. P. & Broitman, S. A. Beneficial effects of post-transfusional hepatitis in acute myelogenous leukemia may be mediated by lipopolysaccharides, tumor necrosis factor alpha and interferon gamma. Leukemia 6, 1036-1042 (1992).
67 Abel, U. et al. Common infections in the history of cancer patients and controls. Journal of cancer research and clinical oncology 117, 339-344 (1991).
68 Mastrangelo, G., Fadda, E. & Milan, G. Cancer increased after a reduction of infections in the first half of this century in Italy: etiologic and preventive implications. Eur J Epidemiol 14, 749-754 (1998).
69 Albonico, H. U., Braker, H. U. & Husler, J. Febrile infectious childhood diseases in the history of cancer patients and matched controls. Medical hypotheses 51, 315-320 (1998).
70 Maurer, S. & Koelmel, K. F. Spontaneous regression of advanced malignant melanoma. Onkologie 21, 14-18 (1998).
71 Schlehofer, B. et al. Role of medical history in brain tumour development. Results from the international adult brain tumour study. Int J Cancer 82, 155-160 (1999).
72 Koelmel, K. F. et al. Infections and melanoma risk: Results of a multicenter EORTC case study. Melanoma Res 9, 511-519 (1999).
73 Stewart, B. W. & Kleihues, P. World Cancer report. (World Health Organization, IARC Press, 2003).
74 Koelmel, K. F. et al. Prior immunisation of patients with malignant melanoma with vaccinia or BCG is associated with better survival. An

REFERENCES

European Organization for Research and Treatment of Cancer cohort study on 542 patients. Eur J Cancer 41, 118-125 (2005).

75 Mastrangelo, G. et al. Lung cancer risk: effect of dairy farming and the consequence of removing that occupational exposure. American journal of epidemiology 161, 1037-1046, doi:10.1093/aje/kwi138 (2005).

76 Jeys, L. M., Grimer, R. J., Carter, S. R., Tillman, R. M. & Abudu, A. Post operative infection and increased survival in osteosarcoma patients: are they associated? Annals of surgical oncology 14, 2887-2895, doi:10.1245/s10434-007-9483-8 (2007).

77 Kim, S. Y. et al. The influence of infection early after allogeneic stem cell transplantation on the risk of leukemic relapse and graft-versus-host disease. American journal of hematology 83, 784-788, doi:10.1002/ajh.21227 (2008).

78 Rudant, J. et al. Childhood acute leukemia, early common infections, and allergy: The ESCALE Study. American journal of epidemiology 172, 1015-1027, doi:10.1093/aje/kwq233 (2010).

79 Urayama, K. Y., Buffler, P. A., Gallagher, E. R., Ayoob, J. M. & Ma, X. A meta-analysis of the association between day-care attendance and childhood acute lymphoblastic leukaemia. International journal of epidemiology 39, 718-732, doi:10.1093/ije/dyp378 (2010).

80 Urayama, K. Y. et al. Early life exposure to infections and risk of childhood acute lymphoblastic leukemia. Int J Cancer 128, 1632-1643, doi:10.1002/ijc.25752 (2011).

81 Rudant, J. et al. Childhood Hodgkin's lymphoma, non-Hodgkin's lymphoma and factors related to the immune system: the Escale Study (SFCE). Int J Cancer 129, 2236-2247, doi:10.1002/ijc.25862 (2011).

82 Chilvers, C., Johnson, B., Leach, S., Taylor, C. & Vigar, E. The common cold, allergy, and cancer. Br J Cancer 54, 123-126 (1986).

83 Cardwell, C. R., McKinney, P. A., Patterson, C. C. & Murray, L. J. Infections in early life and childhood leukaemia risk: a UK case-control study of general practitioner records. Br J Cancer 99, 1529-1533, doi:10.1038/sj.bjc.6604696 (2008).

84 Hoffmann, C., Rosenberger, A., Troger, W. & Buhring, M. Childhood diseases, infectious diseases, and fever as potential risk factors for cancer ? Forsch Komplementarmed Klass Naturheilkd 9, 324-330 (2002).

85 Hobohm, U. Fever therapy revisited. Br J Cancer 92, 421-425, doi:10.1038/sj.bjc.6602386 (2005).

86 Hobohm, U., Stanford, J. L. & Grange, J. M. Pathogen-associated molecular pattern in cancer immunotherapy. Critical reviews in immunology 28, 95-107 (2008).

87   Tanchou, S. (ed Germer Bailliere) (Paris, 1844).
88   Reinhard, E. H., Good, J. T. & Martin, E. Chemotherapy of malignant neoplastic diseases - abstract of discussion, with statement by MJ Shear, Bethesda. JAMA 142, 383 (1950).
89   Morton, D. L. et al. BCG immunotherapy of malignant melanoma: summary of a seven-year experience. Annals of surgery 180, 635-643 (1974).
90   National Cancer Institute Seer cancer statistics review, <http://seer.cancer.gov/csr/1975_2004/results_merged/topic_historical_mort_trends.pdf> (2005).
91   Mantovani, A. & Pierotti, M. A. Cancer and inflammation: A complex relationship. Cancer Lett 267, 180-181 (2008).
92   Eberhardt, M. V., Lee, C. Y. & Liu, R. H. Antioxidant activity of fresh apples. Nature 405, 903-904, doi:10.1038/35016151 (2000).
93   Krieg, A. M. Toll-like receptor 9 (TLR9) agonists in the treatment of cancer. Oncogene 27, 161-167 (2008).
94   Pardoll, D. M. Therapeutic vaccination for cancer. Clinical immunology (Orlando, Fla 95, S44-62 (2000).
95   Matzinger, P. The danger model: a renewed sense of self. Science 296, 301-305 (2002).
96   Willemze, R. Primary cutaneous B-cell lymphoma: classification and treatment. Current opinion in oncology 18, 425-431, doi:10.1097/01.cco.0000239879.31463.42 (2006).
97   Orange, M. et al. Durable Regression of primary Cutaneous B-Cell Lymphoma Following Fever-inducing Mistletoe treatment: two Case Reports. Global Adv Health Med. 1, 18-25 (2012).
98   Kaczanowska, S., Joseph, A. M. & Davila, E. TLR agonists: our best frenemy in cancer immunotherapy. J Leukoc Biol 93, 847-863, doi:10.1189/jlb.1012501 (2013).
99   Sauer, R., Creeze, H., Hulshof, M., Issels, R. & Ott, O. Concerning the final report "Hyperthermia: a systematic review" of the Ludwig Boltzmann Institute for Health Technology Assessment, Vienna, March 2010. Strahlentherapie und Onkologie : Organ der Deutschen Rontgengesellschaft ... [et al] 188, 209-213, doi:10.1007/s00066-012-0072-9 (2012).
100 Wessalowski, R. et al. Regional deep hyperthermia for salvage treatment of children and adolescents with refractory or recurrent non-testicular malignant germ-cell tumours: an open-label, non-randomised, single-institution, phase 2 study. The lancet oncology 14, 843-852, doi:10.1016/S1470-2045(13)70271-7 (2013).

REFERENCES

101 Orange, M., Fonseca, M., Lace, A., Laue, H. B. & Geider, S. Durable tumour responses following primary high dose induction with mistletoe extracts: Two case reports. European Journal of Integrative Medicine 2, 63-69 (2010).

102 Hobohm, U. Toward general prophylactic cancer vaccination. Bioessays 31, 1071-1079, doi:10.1002/bies.200900025 (2009).

103 Mandel, A. L. & Breslin, P. A. High endogenous salivary amylase activity is associated with improved glycemic homeostasis following starch ingestion in adults. The Journal of nutrition 142, 853-858, doi:10.3945/jn.111.156984 (2012).

104 Young, E. Deprive yourself. New Scientist, 47-49 (2012).

105 Ruskin, D. N. & Masino, S. A. The nervous system and metabolic dysregulation: emerging evidence converges on ketogenic diet therapy. Frontiers in neuroscience 6, 33, doi:10.3389/fnins.2012.00033 (2012).

106 Lee, C. et al. Fasting cycles retard growth of tumors and sensitize a range of cancer cell types to chemotherapy. Science translational medicine 4, 124ra127, doi:10.1126/scitranslmed.3003293 (2012).

107 Safdie, F. et al. Fasting enhances the response of glioma to chemo- and radiotherapy. PloS one 7, e44603, doi:10.1371/journal.pone.0044603 (2012).

108 Goodwin, P. J. et al. Fasting insulin and outcome in early-stage breast cancer: results of a prospective cohort study. J Clin Oncol 20, 42-51 (2002).

109 Vigneri, P., Frasca, F., Sciacca, L., Pandini, G. & Vigneri, R. Diabetes and cancer. Endocrine-related cancer 16, 1103-1123, doi:10.1677/ERC-09-0087 (2009).

110 Templeton, A. et al. Prognostic role of neutrophil-to-lymphocyte ratio in solid tumors: a systematic review and meta-analysis. J Natl Canc Inst 106(6), 2014

111 Kox et al. Voluntary activation of the sympathetic nervous system and attenuation of the innate immune response in humans. PNAS 111(20), 7379-7384 (2014)

112 Mantovani *et al*. Cancer-related inflammation. *Nature* **454** (7203), 436-444 (2008)

Although informations in this essay have been collected with great care, no liability is assumed for any detail specified.

Further information can be found at www.fevertherapy.eu

I thank you, my vivid family.
Heinz-Uwe Hobohm, December 2015

The autor

Heinz-Uwe Hobohm studied chemistry and biology in Hamburg and Bremen, Germany and obtained a PhD in cell biology at the University of Bremen. As a post-doctoral scientist he worked at the European Molecular Biology Laboratory (EMBL) in Heidelberg in bioinformatics and as a research scientist at Hoffman-La Roche in Basel. Presently he is Professor in the bioinformatics department at the University of Applied Sciences in Giessen, Germany.

He published several books and articles on cancer.

www.ingramcontent.com/pod-product-compliance
Lightning Source LLC
Chambersburg PA
CBHW031423210526
45464CB00005B/2022